THE CAVES
BEYOND

SPELEOLOGIA

A series of books of general and scientific interest on caving and karst research.

Cave Conservation

Caves are fragile in many ways. Their features take hundreds of thousands of years to form. Cave animals such as blind fish are rare, and always live in precarious ecological balance in their underground environments. Cave features and cave life can be destroyed unknowingly by people who enter caves without informing themselves about cave conservation. Great, irreparable damage has been done by people who take stalactites and other flowstone features from caves, and who disturb cave life such as bats, particularly in winter when they are hibernating. Caves are wonderful places for scientific research and recreational adventure, but before you enter a cave, we urge you first to learn about careful caving by contacting the National Speleological Society, 1 Cave Avenue, Huntsville, Alabama 35810.

THE CAVES BEYOND

The story of the Floyd Collins' Crystal Cave exploration

Joe Lawrence, Jr.
Roger W. Brucker

FOR THE NATIONAL SPELEOLOGICAL SOCIETY

*With a New Introduction by Roger W. Brucker
and an Index by Joan Brucker*

Zephyrus Press, Inc. • *Teaneck*

Library of Congress Cataloging in Publication Data

Lawrence, Joseph, 1924-
  The caves beyond.

  (Speleologia series)
  Reprint of the 1955 ed. published by Funk &
Wagnalls, New York.
  Bibliography: p.
  Includes index.
  1. Floyd Collins' Crystal Cave. I. Brucker,
Roger W., joint author. II. Floyd Collins' Crystal
Cave Exploration, 1954. III. National Speleological
Society. IV. Title.
GB605.K4L3 1975   551.4'4   75-34060
ISBN 0-914264-17-6
ISBN 0-914264-18-4 pbk.

Zephyrus Press, Inc., 417 Maitland Avenue, Teaneck, NJ 07666.

Printed in the United States of America by Braun-Brumfield, Inc.

CONTENTS

Part III

ANOTHER

BARRIER AND CAVES BEYOND

by Joe Lawrence, Jr.

INTRODUCTION TO THE NEW EDITION

More than 20 years of hindsight leaves *The Caves Beyond* standing on its merits as a narrative about the National Speleological Society's week-long expedition in Floyd Collins' Crystal Cave in 1954. It was the first American cave story of its type. Although only 10,000 copies were printed, it has been read widely around the world. In 1975 the going rate for a mint condition copy of the original book was $30. And many cavers today trace an early desire to explore caves to reading *The Caves Beyond.*

Part of the reason for its popularity is its technical content. The 1954 expedition used new techniques. The fact that some of them did not work very well was acknowledged by the 1954 C-3 team. But the ideas were retested by cavers in other situations; some were modified, others junked. Caving is much more than technique, however, and the second reason for the book's appeal is its deep dimension of human adventure. Often quiet, but seldom bored, the expedition members felt excitement as well as frustration, and the book conveys both.

The most frequently asked questions by readers over the years are: What is happening now in Crystal Cave? What are Joe Lawrence and Roger Brucker doing now? Did they ever find more cave at . . . ?

What is happening now? Exploration and survey in the great caves of Mammoth Cave National Park have continued almost without interruption since 1954. Four large caves in Flint Ridge—Crystal, Unknown, Salts, and Colossal—have been connected and enlarged. The result was that as of September 9, 1972, one could enter Crystal Cave and traverse nearly 87 miles of surveyed passages. And by the following day, September 10, the mileage had grown to more than 144 miles through a connection to Mammoth Cave. In other words, the Flint Mammoth Cave System is the longest cave known to man. In a typical recent year, three times as many explorers entered the cave, and they found and mapped three times as much as in 1954. In March, 1975, passage length totaled 169.2 miles, and new surveys have been adding more than a mile a month.

What about Lawrence and Brucker? Joe resides in Houston, Texas, where he is vice president of Datac Company, a manufacturer of traffic light controls. I live in Yellow Springs, Ohio, where I am president of Odiorne Industrial Advertising, Inc. Both of us retain an active interest in caves, but Joe has not gone underground for several years. I continue to be centrally involved in the exploration of the caves of Mammoth Cave National Park.

Did they find more cave? Yes, starting with the first trip into Crystal Cave after the February, 1954, expedition. On that and succeeding trips the discoveries ranged beyond the Bogardus Waterfall Trail. In 1955, explorers stretched the cave in all directions: south Left of the Trap, west into the heart of Flint Ridge, and north to a siphon a few hundred feet from the Green River.

Before recounting subsequent events, there are a few old accounts to settle. It is time to tell some of the untold stories of these 1954 days.

A most dramatic struggle was the conflict that gave birth to the C-3 expedition. Commercial cave rivalries in Kentucky go way back. In 1913, E.-A. Martel, writing in French in *Spelunca,* deplored the spirit of secrecy and obstructionism with which the owners of Mammoth Cave guarded their maps. Twenty years earlier the claimants of Colossal Cave engaged in litigation, one even digging a separate "back-door" entrance, all for commercial gain. The New Entrance to Mammoth Cave was opened by an entrepreneur and the whole south end of the cave shown separately in the 1920s—as a result of a bootleg survey and a determination to crack the monopoly-hold of Mammoth Cave on tourist business.

Floyd Collins himself was an indirect victim of the struggle. When he died in Sand Cave he was trying to gain access to a cave farther south than that reachable by the new entrances, at a better location on the road the tourist traveled. Whether Collins was looking for a back door to Crystal Cave, or just a good business opportunity, is beside the point. The point is, in that part of Kentucky then, as now, caves were a livelihood, not a hobby. If your competitors cut you off, you could starve! Collins well knew the hardship of operating a remote cave off in the far corner of Flint Ridge.

After Collins' death, the pattern of competition settled into a 20-year keen rivalry. The goal was to steer tourists to your cave and away from competitors' caves. Roadside soliciting stands sprang up, staffed by smooth-talking men and women. They flagged down cars and promised the innocent occupants the cave trip of a lifetime, dropping the information that nearby caves were devoid of pretty formations, or even closed for repairs.

In 1926, the movement to incorporate Mammoth Cave into a national park grew from an idea to the organization of a Mammoth Cave National Park Association to promote this purpose. By law, Congress could accept land for a national park only as a gift. The mechanism was for a state-backed group to buy up land—nearly 50,000 acres of it. Proceeds from the Association's operation of Mammoth Cave were to pay back the costs of land acquisition in excess of citizens' donations. In 1941, Mammoth Cave officially became a national park, surrounding two islands of private holdings—Floyd Collins' Crystal Cave and Great Onyx Cave. One might expect commercial rivalry to diminish with the federal presence, but the intensity increased. The owners of Crystal Cave and Great Onyx Cave moved to counter what they saw as official Park Service attempts to steer tourists away from their caves. They dressed their roadside solicitors in kakhi uniforms. They put up flagpoles. One ticket stand for Diamond Caverns, a competitor outside the Park, featured artfully stationed green pickup trucks under a canopy to resemble every tourist's idea of what the official Mammoth Cave information office must look like.

The owners and operators of the commercial caves began to feel another kind of systematic pressure. The owners of Crystal Cave and Great Onyx Cave had refused to accept the low prices offered for their going commercial concerns. It seemed to them that the government was now engaged in a deliberate attempt to reduce the value of their properties by every means possible, to force a cheap sale at distress prices. Federal officials seemed blind to the idea that there might be enough business to go around.

Evidence for this pattern was not hard to see. Both caves were reached by roads that were maintained through Park lands by court order. Beyond the macadam at Mammoth Cave, the road became a rutted clay trail. Occasionally a scraper would push some gravel around and smooth over the ruts, provided the cave owners protested enough. But many tourists bent on seeing one or both of the caves concluded that they had made a mistake—this trail couldn't possibly be the road to a commercial cave!

Jim Dyer, manager of Crystal Cave, heard reports that customers were being steered away. He decided to confront an individual at Mammoth Cave he felt was responsible. For his trouble he was beaten up and threatened with death the next time he interfered.

A more overt form of harassment was a lawsuit by the National Park Service to condemn the Crystal Cave gravel pit. It was the only economical source of road gravel for maintaining the private portion of the road to Crystal Cave. Crystal Cave owners were able to prove to the satisfaction of the court that they were entitled to dig the gravel until they were paid a fair price for the land in question.

Competitive pressures took other forms, such as ticket price wars, threats and pressures to lure away or drive off cave guides at the private caves, and an arrogant pronouncement to the owners of Crystal Cave that their commercial days were numbered. They had better sell while they still could, because there were plenty of other tricks in the bag. The message was clear: The government wanted Crystal Cave at the lowest possible price, and seemed prepared to stop at nothing to destroy the owners' business.

In 1952, Bill Austin became manager of Crystal Cave. Both he and his uncle, Dr. E. Robert Pohl, manager of Mammoth Onyx Cave, were responsible to Mrs. H. B. Thomas, owner of the cave properties. They did everything within their power to counter the government squeeze by increasing the cave's business through better marketing. Bill Austin became a superb color photographer: His slides and pictures made it possible for people to see the beautiful formations and scenes at Crystal Cave in new contexts. Their new billboards extended far along the roads toward Louisville and Nashville. But as business increased, the government negotiators offered less money for the land. In private conversations they revealed their contempt for this "insignificant" cave.

The strategy of Austin and Pohl may never be known for certain, but from their actions it can be guessed. As early as 1947 they began to invite cave explorers to look carefully at the cave, and they began to plan more extensive tours. Austin, and Dyer before him, had taken friends into the lower levels. The cavers were always impressed by the wonders they saw, not the least of which was the extent of the labyrinth. Austin and Luther Miller had surveyed the Crawlway and some of the lower levels. They knew that it would be possible to sink a shaft from the tourist trail beyond Scotchman's Trap down to the top of X Pit. That would open up a beautiful waterfall trip, with unsurpassed canyon scenery. Austin was working on the shaft when Joe Lawrence and his friends arrived.

The idea may have grown in the managers' minds, partly as a result of Lawrence's visit (which is where *The Caves Beyond* begins), that by inviting an expedition of the National Speleological Society to Crystal Cave—which Dr. Pohl as a geologist wished to do anyway for the sake of scientific research—two outcomes might be expected. First, an accurate description of the true extent and scientific value of Crystal Cave might make a strong case that the cave was in fact significant. Second, there would be publicity attending a really big expedition. Publicity would draw more tourists. Tourist dollars would swell the revenue and force up the takeover price. If inevitably we will be bought out, they probably reasoned, we might as well get a fair price. In April, 1953, Dyer and Austin hosted the week of field activities preceding the NSS convention in Louisville, and showed speleologists Crystal Cave. Many who were to take part in the C-3 expedition first became acquainted with the cave and its managers at that time.

Thus, the first untold story: Had the government not tried to squeeze out Crystal Cave at a ridiculously low price, there might never have been a C-3 expedition. And while the expedition planners may have surmised that Austin and Pohl were hoping for favorable publicity to help their cause, Lawrence and company had entirely different reasons for the undertakings. They mainly wanted to explore the cave.

As the book recounts, Phil Harsham was one of the first reporters on the scene. Austin told me many years after the expedition that Harsham's secret purpose was to expose what he had been led to believe was a hoax. It was thought that Crystal Cave was not big enough to hold an expedition, that the whole thing was a monster publicity stunt, which a clever investigative reporter would soon be able to see through. Of course, I knew none of this at the time. Harsham was a good investigative reporter, and he soon saw that there was a lot more to the cave and the expedition than mere puffery. He, more than any other reporter, was responsible for the lavish press coverage that resulted in front page stories in hundreds of newspapers.

Speaking of reporters, before going on to more untold stories of those 1954 days, I should like to print what was unprintable then. In the book we reported Bob Halmi's exclamations as "son-of-a-gun." He really said "son-of-a-bitch," milking every consonant for the highest possible dramatic effect.

A second untold story is hinted at in the next-to-the-last paragraph in Chapter 15. It describes our survey near the Bottomless Pit, but omits the fact that we tied the survey to the Pit itself. The significance of including Bottomless Pit was that it gave us the necessary key to draft a map of the entire cave.

Roy Charlton, Earl Thierry, and I had been frustrated not knowing where the cave passages lay in relation to the upper levels and to the surface. We concluded that Bill Austin would not or could not make his own map available to us. So we hit upon a desperate gamble. The tourist folder map of the upper level of Crystal Cave looked like a genuine survey. It even contained the legend: *This map was prepared from an accurate instrumental survey.* But it bore no north arrow nor any scale. Was it accurate?

We decided to do three things: Earl would make a series of compass sightings in the upper levels to try to orient the map. Roy would pace off some recognizable segments of the mapped passages to try to learn the scale. And I would carry the 3-C survey to the brink of Bottomless Pit—the last landmark on the tourist folder map. We agreed to try to fit everything together after the expedition.

To make a long story short, the crude composite map resulted in a pretty good understanding of the cave. It formed the basis of the isometric view of the cave shown on pp. 56-57. Bill Austin was dismayed at our sneaky tactics, but admitted that there was no other way we could have obtained a map to publish in the book. The reason for secrecy, we learned later, was to prevent the National Park Service from knowing where the cave extended under their property. If they knew, they might have forbidden further exploration.

Another untold story is how *The Caves Beyond* was written. Funk & Wagnalls, who had paid the Society a $2000 advance to write a book about the C-3 expedition, demanded the manuscript. The Society, through its public relations vice-president, had contracted with a man to ghost-write the book. The ghost writer had gathered together all the logs and transcribed tapes from the expedition. He had solicited personal memoirs from expedition members. I had sent him nearly 100 pages of recollections. But months had gone by and there was no book. The publisher wanted action: Produce, or return the money.

Joe Lawrence, Howard Sloane, and other NSS officers burned up the telephone wires trying to figure out what to do. Lawrence suggested contacting me. Would I be willing to help write a book about the C-3 expedition?

The main question was, *could* I? We discussed it on the phone. I would try to get special leave from the Air Force to go to New York. Unfortunately, I had used up all the leave I was entitled to, and there was no one in authority in my unit who was sympathetic. After inquiry, I found that if I were to have a minor elective operation (which I needed), I could expect two weeks of convalescent leave. The surgeon himself, who was writing a medical book, understood the situation perfectly.

"We'll just schedule a pilonidal sinusectomy. You'll be in the hospital three or four days, and then you'll go on convalescent leave," he said. In I went under the knife. But when I came to, my surgeon had vanished. He had picked that time for a leave of his own. Then, horror of horrors, my records were lost for days, dragging on for a week and a half. Each inquiry brought a cheery, "Oh, we'll find them!" from the new doctor or the head nurse.

"You've made a splendid recovery," said the doctor one day, "so I'm cutting orders to send you right back to your unit."

"But what about the convalescent leave I was promised?" I protested.

"Not necessary," said the doctor.

I thought fast. "I'm a motion picture script writer in my unit," I said, "and I have to spend all day pounding a typewriter. You may think my operation is healed, but my seat is so sore I am positive I could not concentrate on film writing for at least two weeks, or maybe longer."

The doctor reluctantly restored orders for my leave. That evening I saw the lights of Ohio and Pennsylvania slip by under the wing of the Constellation as I headed for a rendezvous with the editor at Funk & Wagnalls. Ida Sawtelle and Howard Sloane met me at the airport. Howard was a generous-spirited New Jersey caver, a friend from my days in New York in 1952. I had met Ida also in 1952. I had come to like and admire her before and during the C-3 expedition. She occupied a three-story house in Brooklyn, headquarters of her dog training and boarding business. The house had a large attic that would be my home for two weeks. Had it not been for Ida Sawtelle's hospitality and good humor, there probably would not have been a book called *The Caves Beyond*.

The following morning, Friday, I went to the Manhattan offices of Funk & Wagnalls, and there met Bill Sloane, the editor. He said he had read the material I had sent to the ghost writer and he liked it. I asked him the question uppermost in my mind: "Where do you begin a book?"

I was prepared for a scornful question in return about how could I hope to write a book if I didn't know that. Instead Bill Sloane earnestly answered the question. "You ought to begin a book at the first point at which the action that follows becomes inevitable." He explained that if you start too soon, you bore the reader. If you start too late, the reader would feel cheated. That was the best writing advice I ever received.

I told Sloane that Joe Lawrence was joining me on Saturday, and that on Monday we would deliver the outline for the book, plus the first and last chapters. He looked skeptical but approved the timetable. I thought the assignment would be challenging but not impossible, because Joe was planning to spend the two weeks with me in Ida's attic.

Joe arrived. I told him of the meeting with Sloane, and showed him the workspace, the attic with several long tables, typewriters, and plenty of room. Some Met Grotto members would come in to type every evening. "Fine," Joe said. "There are only two problems. I can't be with you except on weekends because I couldn't get off work. Also, I'm supposed to speak tonight at a meeting of the NSS Northern Regional Organization."

I felt a trapdoor open under me, a moment of panic. "You surely don't mean tonight," I said.

"Yes, tonight, upstate near Albany." He had to start driving right then.

I experienced waves of mixed feelings, anger at being let down, irritation that we could not spend at least the first weekend together, and a curious kind of rapture that *I would write the book*! Panic gave way to the elation that can come from sudden and unexpected responsibility. I have come to know this emotion from cave exploring, and to seek it. I feel great pleasure and excitement when I "wrest leadership" from another, move just a bit uncertainly into the unknown, and find big cave. And I have deliberately reversed the roles, turning over the lead with great personal anguish, to give another the ultimate experience of finding big cave. Relinquishing leadership at precisely the right time is a very powerful way to produce in others the strong, almost fanatic commitment necessary to conquer the Flint Ridge Cave System. Maybe Joe and the others had something like that in mind. But I did not wrest the lead here. Up in Ida's attic, it was dumped in my lap.

The practical thing to do was to close the writing factory for the weekend to join Joe on his upstate trip. We were able to work out a comprehensive outline of the book during the journey, and we made the decision that the opening and closing chapters would be written from Joe's point of view. The central section would be from my point of view. I would write the entire manuscript. Joe would edit it on weekends. Joe would write the photograph captions. On our return on Sunday, we completed the outline, and Joe went back home to Philadephia.

On Monday afternoon I delivered our outline and the first and last chapters to an amazed Funk & Wagnalls editor. Then the pattern evolved: Write until about 2 A.M., read unrelated books for an hour or two, then sleep until 10 A.M. Write at a furious pace until about 5 P.M. Then the typists would arrive. I would edit what they had typed the day before while they typed the new material. At 10 P.M. the typists would leave, and I would continue writing until 2 A.M.

With a comprehensive outline it was possible to skip around. Every few days I would take some of the finished manuscript into Manhattan. Those days were the most productive of my life, and I still relive the excitement when I think of them. At the spring convention of the NSS in 1955, Joe and I were awarded life memberships in the Society for having produced the book. Life memberships were $75 then. The book itself was the real reward for that effort, but the honor was genuinely appreciated.

Reviewers generally found *The Caves Beyond* to be "exciting . . . a dramatic record . . . the suspense of an adventure story." My favorite comment is from one of the few adverse notices, a review in the *New Yorker* of September 17, 1955:

One would expect an account of the National Speleological Society's exploration of Crystal Cave, in Kentucky, in 1954, to be pretty stimulating, but the expedition was so well equipped and so self-conscious (its telephone conversations were all taped and many are printed here verbatim) relative to the little it discovered and the mildness of its adventures that this book sometimes suggests a team of African explorers braving the Central Park Zoo.

The part about the Central Park Zoo was pretty funny, but I shared the reviewer's opinion that we had been over-equipped. It echoed my personal resolve, as an expedition member, to discover more in the future with less burdening structure than we had on C-3.

This is not the place to cover all of the exploration and discoveries that spanned 1954-72 in Crystal Cave and Flint Ridge. That is the subject of another book, *The Longest Cave,* by Roger W. Brucker and Richard A. Watson, scheduled for publication by Knopf in 1976. I will try to compress the bare bones of it into a few paragraphs.

Toward the fall of 1954, Phil Smith and Jim Dyer had the rest of us working to field an expedition that would remedy the shortcomings in organization and technique we felt were inherent in the C-3 expedition. Because this was our show, we called the shots. Our target was outside the Park, James Cave, chosen for its deep pit which offered attainable objectives of a difficult crossing and descent. We attained most of the goals we set, plotting the map as the exploration progressed and keeping in touch with the surface by means of a lightweight telephone rig, an improvement that newcomer David Jones brought. We probed the depths of the pit but never really explored the other side of it. And we wrote and provided photographs for some publicity that paid the bills for the expedition.

1955 was a year of assimilation. Page proofs of *The Caves Beyond* were checked. Flint Ridge cavers began in May to probe the intriguing area called Left of the Trap, and Black Avenue near the Overlook. Joe went to the cave with his wife and daughters; he explored the downstream upper level of Eyeless Fish Trail, ending in a sand fill a few hundred feet from Green River. My wife took a trip to the Lost Passage, and got stuck in the Hole in the Ceiling.

We simplified caving techniques. Instead of an assortment of heartburn-producing food such as candy bars and Spanish rice, most of our cavers standardized on small cans of boned chicken, fruit, and date-nut roll. This combination made a satisfying, digestible meal. As the work went on, Austin had given cavers the use of the Crystal Cave guides' house for bunk space. During the hot June expedition, individual groups had cooked separately and food preparation took too long. We decided to institute a mandatory communal kitchen. It was one of the wisest camp management moves we ever made. Not only did it cut down enormously on the man-hours and bother of cooking meals, but also the group meals provided the setting for cementing together an already close community.

We convinced the Central Ohio Grotto that we ought to build a cookhouse near the bunkhouse. Austin gave permission, and Dave Huber and Burnell Ehman drew plans. By fall of 1955, we had a new plywood kitchen, attractively white inside with rough-hewn 2 by 4 oak beams above, and painted green outside. A rusty old cast-iron "Worthy Washington" stove would supply heat for the next 18 winters. (This one-room building became known as the Spelee Hut, and was to be our field headquarters until 1972.)

At Crystal Cave, on two trips we made a breakthrough by pushing beyond Bogus Bogardus (read *The Longest Cave*), over some lily pad formations into a virgin crawlway that led to a fantastic vertical shaft 150 feet high and 40 feet in diameter. It was called the Overlook. A drain passage led to a major river passage containing cave blindfish. Throughout 1955 we continued to advance in this area.

Toward the end of 1955 we were able to connect Crystal Cave with Unknown Cave, located close to the center of Flint Ridge. In our announcement of this discovery, made at the December 1955 meeting of the American Association for the Advancement of Science, we said we had the world's longest cave. Our maps showed 23 miles of passageway in the Flint Ridge, and we knew of 14 more miles we could survey in a hurry if we got more manpower. We were stunned when the news came from Switzerland that Hölloch (Hell Hole Cave) in that country was even longer. We wrote to Prof. Dr. Alfred Bögli and verified the truth of the news report, and struck up a friendship that has continued ever since.

In May, 1956, Bill Austin completed construction of a new entrance to the cave, a "back door" that put us into the farthest reaches of Crystal Cave, and made accessible to us the whole of Unknown Cave. We spent the next four and a half years exploring and surveying passages that spread out under lands west toward Great Onyx Cave, south toward Colossal Cave, and southeast toward Salts Cave. We even found a passage that took us part way under Houchins Valley, the deep barrier that separates Flint Ridge from Mammoth Cave Ridge.

In 1959 Phil Smith negotiated an agreement with the National Park Service. Signed on the eve of the sale of Floyd Collins' Crystal Cave to the government, the agreement permitted exploration in all Flint Ridge caves. With the sale, the National Park Service attitude changed to one of friendly interest.

In August, 1960, we found a connection between Colossal Cave and Salts Cave, although we had been trying to hook Colossal Cave to the Crystal/Unknown part of the System. The following August we found a connection between Crystal/Unknown and Colossal/Salts. To simplify the nomenclature, we called the system the Flint Ridge Cave System. By 1961, our cave was in a neck-and-neck race with Hölloch as to which was the world's longest. Through the 1960s our expeditions took us closer and closer to Mammoth Cave.

What started after the C-3 expedition as a project of the Central Ohio Grotto of the NSS, evolved to become a nationally sponsored NSS project. Eventually we established the Cave Research Foundation (CRF), partly as a vehicle to continue the work in Flint Ridge, and partly from the growing understanding that we had the rare opportunity to have a long-lasting effect on world speleology through a specialized organization.

Phil Smith served as the first CRF president, followed by Red Watson, Joe Davidson, Stan Sides, and now me.

In 1967, we began to supply information to the study team putting together the Mammoth Cave National Park Master Plan. At this time we deliberately stopped looking for a connection to Mammoth Cave for political reasons. Some members of the study team revealed their view that Mammoth Cave was not a wilderness, but that it had been spoiled by its development. They believed that the Flint Ridge Cave System represented the only wilderness cave resource in the Park. We feared that if we connected the two, a case might be made for spoilage-by-association: Flint Ridge could not be wilderness if it was an integral part of the spoiled cave. We knew that in fact not only the Flint Ridge Cave System, but also most of Mammoth Cave, is underground wilderness. Over the years, more and more Park Service officials agreed with us.

The early Park study team showed a blind spot also by speculating that perhaps there was no large cave system under Joppa Ridge, flanking Mammoth Cave Ridge to the west. We knew that the same geological and hydrological conditions prevail beneath Joppa Ridge as do beneath Flint Ridge and Mammoth Cave Ridge. There was circumstantial evidence that Joppa contained large caves. But it had to be shown in the ground. So if politics kept us away from Mammoth Cave Ridge, politics threw us into Joppa Ridge.

We had already mapped parts of Proctor Cave and Long's Cave, good-sized trunk passages, in Joppa Ridge. So we looked for more. Gordon Smith and his wife-to-be, Judy Edmonds, spent many weekends tramping over the edges of Joppa Ridge. One weekend Gordon found an entrance to a pit. Near the bottom, scratched on the wall, was: T. E. LEE 1876. CRF survey teams descended a tight canyon and found a magnificent trunk passageway a mile and a half long, 60 to 100 feet wide and 30 feet high in spots. We called it Lee Cave, and soon Gary Eller found another entrance to it. By 1972, we had mapped about seven and a half miles of passages in Lee Cave. The Joppa Ridge Cave System could no longer be ignored. And at this point we said politics be damned. With big cave systems on either side of Mammoth Cave, we decided it was time to begin in the middle.

When work began in 1969 in Mammoth Cave, it was with a bang. We started mapping and exploring Mammoth Cave, under our agreement with the National Park Service. Red Watson, editor of the Zephyrus Press "Speleologia" series, fielded six parties in Mammoth on the first day, mapping almost two miles of passageway. We were so preoccupied that nobody noticed until later that in March, 1969, we finally surpassed Hölloch in length. The Flint Ridge Cave System has been the longest in the world since that time.

Then, on an expedition in May, 1972, some of us began to see that we were very close to connecting the Flint Ridge Cave System with Mammoth Cave. There were many trips, some ending in frustration, but on August 30, 1972, Patricia Crowther led Tom Brucker (my son), John Wilcox, and Richard Zopf on a record-breaking trip into a newly discovered river under Houchins Valley. Along that passage they found scratched on the wall the name of a former Mammoth Cave guide. On September 10, Wilcox, Zopf, Patricia Crowther, Gary Eller, Steve Wells, and Park Ranger Cleve Pinnix completed the first portal-to-portal trip, tying the survey from Flint Ridge through the new river into Mammoth Cave's Cascade Hall. When they left Mammoth Cave in the dark of early morning they knew they had climbed the Everest of speleology.

The newly integrated Flint Mammoth Cave System totaled 144.4 miles at that time, although subsequent trips have added 25 miles. We shot past the 150 mile mark, a legendary, fanciful figure that had been cited for publicity purposes as Mammoth Cave's length for a hundred or more years. After the connection we devoted about half of our efforts to Mammoth Cave, a fourth to Flint Ridge, and the other fourth to Joppa Ridge. Maps are now plotted by computer from punched-in survey data. As the pace quickens, the caves show no signs of running low on leads.

The Caves Beyond, apart from being an historic benchmark in the continuing saga of cave exploration in the Kentucky Cave Region, is one of the first American accounts of what it is like to face underground wilderness at its most demanding. Elsewhere there are vivid accounts of individual cave trips, stark terror or fantastic discoveries, and gripping descriptions of fatal descents into deep pits. But the endurance barriers of mile after mile of punishing length in Crystal Cave and the larger Flint Mammoth Cave System are matched nowhere else in the world except in Bögli's Hölloch in Switzerland.

If our adventures seem tame because there has so far been so little violence in our encounters with underground nature, there is nevertheless ample danger. We cavers do not want the terrible adventures that newspaper readers find exciting. What the *New Yorker* reviewer whom I quoted above forgot is that we may be operating on borrowed time. There are lions there and it is always night. The cave can maul or kill in many ways. The degree of danger depends on the competence, experience, and care of the adventurer, and like the best climbers of the great vertical walls of Yosemite, we who explore the longest cave feel confident in our skills. We push ourselves to the limit.

What has the passage of time done to the feelings of exhaustion, malaise, bone-racking weariness, and the heady joy upon entry into that new cave in 1954? Memory sharpens those C-3 images, so overwhelming at the time that *The Caves Beyond* was written. Now—decades later—the feelings have become sharpened also by further experience. We did not know in 1954 that we were savoring the highs and lows of human experience, some of the most intense of our lives, naively perhaps, but purely indeed. We were on the foothills of the Everest of speleology. Some of us made it.

Much is still to be discovered, long trunk passages, spectacular vertical shafts, cave patterns, and more. And more than ever we find delight in the microphenomena, too, the play of light and shadow as an exploring party moves through a glistening wet-walled canyon, the surprise of finding an overlooked crawlway a few feet from an entrance or a well-traveled route, the fascinating interplay of cross-bedding in the limestone walls, the minerals and tiny troglodytes. The lure of the caves in Mammoth Cave National Park is not dimmed in the least, either, by the reexploring of old cave passages. New explorers, like children growing up, experience again the age-old thrill, and old explorers relive the past. If it looks like virgin cave to you, it is virgin. The old cave is ever new.

We started out to explore and to understand the wilderness of caves. And in any wilderness I can imagine, one is confronted by choices. Take these wilderness caves in Mammoth Cave National Park, for example. One can ignore them or explore them. If the choice is to explore, one must further choose degree of commitment, direction, and length of time. Go right or left? Up or down? Continue or retreat? Which will it be? Simple choices. But the Lady or the Tiger—perhaps both together—lies down every passage where one has never been. The challenge of the unknown in caves of this scope beckons to some, and turns others away. It is a challenge greater than any single individual can meet.

Do not doubt that the cave is full of risks, and that the real possibilities of accident require care and prudence. Thus the reason that we require *real* commitment from our coexplorers—and not just casual interest—is that our lives depend on it. With death in the offing, in this serious work, we make some of the deepest and most understanding friendships of our lives. What I am getting out of all this is a full lifetime of learning and adventure, something, I might add, that has been completely and actively shared by my wife and four children, one of whom took some strenuous trips in Crystal Cave before she was born. Two others, Tom and Ellen, became cavers. In the big cave, the lure is always to know and to understand . . . more.

As in any wilderness struggle, one is directly responsible for choices, because the results follow at once. With inexperience in caves, you may blame or credit luck for discoveries or lack of them. With years of experience, you accept the responsibility for outcomes from alternatives knowingly selected. You learn to tolerate the hours and years of frustration, always looking to find the answer around the next bend in the passage. If the passage does not go, wait until next time! There will be a way.

And now a final comment. The people of the Kentucky Cave Region tend to be strong individuals. Self-reliant and tenacious, they are also curious about their caves, and some of America's greatest cave explorers—the black guide Stephen Bishop, the Lee brothers, Floyd Collins, and a score of others—came from their own. Jack Lehrberger and Bill Austin, weekend after weekend in the 1950s, together and alone on trips of up to 26 hours in Flint Ridge caves, set a modern standard for cave exploration that a few others have approached, but which no one will surpass. I could name a few names of those who have followed Bill and Jack, but not here; they know and we know, and that is enough.

Cave exploring also takes self-reliance and tenacity. One must have the will to endure grinding fatigue underground, and on the surface in the Central Kentucky Karst, suffocating heat with chiggers and ticks and the endless calls of whip-poor-wills in the dark summer nights, and wet, freezing cold in the winter. Obdurate, self-assured conviction about one's interests and rights as free citizens of the republic—sometimes expressed in willful political action—seems to be a primary requirement for survival both for cave explorers and for natives of the region. We in CRF have learned persistence and tenacity from the local people and from the caves. Jim Dyer taught us that we had to be as tough as the natives. Bill and Jacque Austin have never let us forget it. Now, after more than 20 years and considerable success, we know that we are in for the long haul.

Roger W. Brucker
July, 1975

BIBLIOGRAPHY OF WRITINGS ABOUT EXPLORATION IN THE FLINT MAMMOTH CAVE SYSTEM

Anonymous. "Explorers Conquer Kentucky's Deepest Cave." *The Columbian Crew*, Vol. 40, No. 9 & 10, 1955, pp. 8-9.

Anonymous. "Explorers Find Major New Cavern in Mammoth Cave National Park." *Louisville Courier-Journal.* April 16, 1969, Sec. 1, p. 1.

Bedford, Bruce L. "The Golden Dream." In *Challenge Underground*, pp. 148-155. Teaneck, New Jersey: Zephyrus Press, 1975.

Blinkhorn, Tom. "Mapping Crystal Cave." *The Cincinnati Pictorial Enquirer*, September 23, 1956, pp. 26-29.

Blakeslee, Alton L. "Crystal Cave Held Biggest in World." *Louisville Courier-Journal*, December 28, 1955, Sec. 2, p. 1.

Brucker, Roger W. "The Death of Floyd Collins." In *Celebrated American Caves*, edited by Charles E. Mohr and Howard N. Sloane, pp. 151-171. New Brunswick, New Jersey: Rutgers University Press, 1955.

Brucker, Roger W. "The Impossible Pit." In *Celebrated American Caves*, edited by Charles E. Mohr and Howard N. Sloane, pp. 78-89. New Brunswick, New Jersey: Rutgers University Press, 1955.

Brucker, Roger W. "Recent Explorations in Floyd Collins' Crystal Cave." *National Speleological Society Bulletin*, Vol. 17, 1955, pp. 42-45.

Brucker, Roger W. "Truncated Cave Passages and Terminal Breakdown in the Central Kentucky Karst." *National Speleological Society Bulletin*, Vol. 28, 1966, pp. 171-178.

Brucker, Roger W. *See also* Rogers, Warren.

Brucker, Roger W. and Denver P. Burns. *The Flint Ridge Cave System* (Map Folio), Washington, D.C.: Cave Research Foundation, 1966. 34 pp.

Brucker, Roger W., David B. Jones, William T. Austin, and Brother G. Nicholas. "New Discovery Yields World's Largest Cave." Paper presented at the American Association for the Advancement of Science Meeting, Atlanta, Ga., December 27, 1955. 5 pp.

Brucker, Roger W., Philip M. Smith, Joe Lawrence, Jr., and David B. Jones. "Some New Approaches to Speleology." Paper presented at the National Speleological Society Convention, Louisville, Ky., April 16, 1955. 11 pp.

Brucker, Roger W. and Richard A. Watson. *The Longest Cave*. New York: Alfred A. Knopf, Inc., forthcoming.

Burman, B. L. "Kentucky's Crazy Cave War." *Colliers*, Vol. 131, 1953, pp. 62-65.

Creason, Joe. "Peace Among the Cave Men?" In *Louisville Courier-Journal Magazine*, February 8, 1959, pp. 5-9.

Crowther, Patricia P. "Into Mammoth Cave the Hard Way." *National Parks & Conservation Magazine*, Vol. 47, No. 1, 1973, pp. 10-15.

Deike, George H., III. *The Development of Caverns of the Mammoth Cave Region*. Pennsylvania State University Ph.D. Dissertation, 1967. 235 pp.

Finkel, Donald. *Answer Back*. New York: Atheneum, 1968. 38 pp.

Folsom, Franklin. "Bigger and Deeper." In *Exploring American Caves*, pp. 153-158. New York: Crown Publishers, 1956.

Freeman, John P. (ed.). *CRF Personnel Manual* (second edition). Columbus, Ohio: Cave Research Foundation, 1975. 109 pp.

Freeman, John P., Gordon L. Smith, Thomas L. Poulson, Patty Jo Watson, and William B. White. "Lee Cave, Mammoth Cave National Park, Kentucky." *NSS Bulletin*, Vol. 35, 1973, pp. 109-125.

Halliday, William R. "The Largest Cave? The Story of Flint Ridge." In *Depths of the Earth*, pp. 333-347. New York: Harper & Row, 1966.

Halmi, Robert. "Crawl Down to Hell." *Eye*, September, 1954, pp. 28-33.

Halmi, Robert. "Discovery of a Bottomless Hole." *True*, Vol. 29, No. 250, 1958, pp. 12-15.

Halmi, Robert. "Report from Underground." *Sports Illustrated*, Vol. 4, No. 1, 1956, pp. 12-16.

Halmi, Robert. "Seven Days in the Hole." *True*, Vol. 34, No. 206, 1954, pp. 40-45, 90-91, 104.

Harsham, Philip. "Diary of a 151-hour Night." *Louisville Courier-Journal Magazine*, August 26, 1954, p. 3.

Heimann, Lee. "Revolt in the Cavern Country." *Louisville Courier-Journal Magazine*, January 29, 1956, p. 1.

Heimann, Lee. "Tri-Level Design." *Louisville Courier-Journal Magazine*, September 30, 1956, pp. 58-59.

Jackson, George F. "The Russell Trall Neville Expedition to Old Salts Cave, Kentucky." *Journal of Spelean History*, Vol. 2, No. 3, 1969, pp. 45-50.

Lawrence, Joe, Jr. and Roger W. Brucker. "The Caves Beyond." *Man's Illustrated*, Vol. 1, No. 5, 1956, pp. 28-31, 58.

Lyons, Richard D. "A Link is Found Between Two Major Cave Systems." *The New York Times*, December 2, 1972, pp. 31, 58.

Metzgar, Hal. "Ladies' Day in a Wild Cave." *Cincinnati Pictorial Enquirer*, March 31, 1957, pp. 26-29.

Norman, Phil. "The Cave Mappers." *Louisville Courier-Journal & Times Magazine*, October 22, 1967, pp. 1, 16-25.

Phinizy, Coles. "A Coon Trap Leads to a Labyrinth." *Sports Illustrated*, Vol. 4, No. 1, 1956, pp. 17, 54-55.

Pohl, E. R. *Vertical Shafts in Limestone Caves.* National Speleological Society Occasional Paper No. 2, 1955. 24 pp.

Rogers, Warren [Roger W. Brucker]. "Assault on Forty Fathom Pit." *Popular Mechanics*, Vol. 106, 1956, pp. 65-69.

Rogers, Warren [Roger W. Brucker]. "Ohio Explorers Conquer Kentucky's Deepest Cave." *Columbus Dispatch Magazine*, October 10, 1954, pp. 1-4, 7-12, 26-27.

Poulson, Thomas L. and William B. White. "The Cave Environment." *Science*, Vol. 165, 1969, pp. 171-181.

Smith, Philip M. "Discovery in Flint Ridge, 1954-1957." *NSS Bulletin*, Vol. 19, 1957, pp. 1-10.

Smith, Philip M. "The Flint Ridge Cave System: 1957-1962." *NSS Bulletin*, Vol. 26, 1964, pp. 17-27.

Smith, Philip M. and Richard A. Watson. "The Development of the Cave Research Foundation." *Studies in Speleology*, Vol. 2, 1970, pp. 81-92.

Quinlan, James F. "Central Kentucky Karst." *Méditeranée*, No. 7, 1970, pp. 235-253.

Watson, Richard A. "Floyd's Coffin." *Inside Earth*, Vol. 1, No. 3, 1974, p. 26.

Watson, Richard A. "Notes on the Philosophy of Caving." *NSS News*, Vol. 24, 1966, pp. 54-58.

Watson, Richard A. and Philip M. Smith. "Underground Wilderness, a Point of View." *International Journal of Environmental Studies*, Vol. 2, 1971, pp. 217-220.

Wells, Stephen G. and David J. DesMarais. "The Flint Mammoth Connection." *NSS News*, Vol. 31, No. 2, 1973, pp. 18-23.

Wheaton, Bob. "Return of the Caveman." *The Columbus Sunday Dispatch Magazine*, August 9, 1953, pp. 10-13.

White, William B., Richard A. Watson, E. Robert Pohl, and Roger W. Brucker. "The Central Kentucky Karst." *Geographical Review*, Vol. 60, 1970, pp. 80-115.

Wilson, Harry. "Following in Floyd Collins' Footsteps." Unpublished manuscript. 7 pp.

FOREWORD TO THE ORIGINAL EDITION

TINY PINPOINTS OF LIGHT push slowly along maddeningly complex passageways that penetrate like some giant spider web for miles in unknown directions, hundreds of feet below the rolling Kentucky landscape. The puny lights cross a pit too deep to illuminate without reinforcement. Later, from those depths, massed lights will show how truly terrifying the traverse was, yet the crossing was made as a matter of course. It had to be made because there were caves beyond.

Pits and canyons, crawlways and chimneys that would have challenged the skill of the most experienced rock climbers were negotiated by men and women from all walks of life—negotiated *in the dark*. In almost every sense this was *mountaineering at night*.

That the thousands of man-hours spent groping through the implacable darkness were accomplished without accident may be credited to kind fortune, good discipline, and superb organization.

Less experienced cavers, even at the peak of condition, would have been running unconscionable risks. But speleology has established a rigid code of safety and exploratory techniques as precise as those of the rock climber or mountaineer. There are few places for the novice in an enterprise where one accident might jeopardize the safety of others. There were many times and places in Floyd Collins' Crystal Cave where it would have been impossible to evacuate the immobilized victim of a serious accident.

Only the expedition's best and most experienced explorers ever went off alone into unknown passageways—and then only when a short time limit for return had been set. Absence beyond that deadline called for immediate investigation. On a few occasions the reinforcements arrived when need for help was critical.

Unclimbed mountains are reconnoitered from every possible angle. Photographs from different points of the compass and from the air reveal vital information about the problems that will be encountered. Not so in the world underground. Those who assault the unknown below the surface must master each new obstacle as it is encountered. Otherwise valuable time is lost —time purchased dearly, at the expenditure of prodigious effort, by scores of supply personnel.

A ton and a half of food and supplies had to be dragged and pushed for thousands of feet, sometimes through crevices ten inches wide, beneath ceilings a foot high, along ledges only inches wide above gaping pits a hundred feet deep. The eight miles of telephone line, consisting of two party lines and 14 phones, were a monument to electrical know-how, to exploring skill and persistence.

The world underground is sparsely inhabited. Too much of its thousands of miles of passageways is devoid of any organic material, dead or alive, to support life. Only where surface streams can carry in organic debris, or in places close to the surface, can the bio-speleologist find concentrations of cave creatures.

Crystal's explorers learned that living hour after hour, day upon day in darkness, without experiencing the familiar, too little appreciated sunrise or the satisfaction of the day-ending sunset, plunged them into a state of unreality. Some persons never can adjust to such an unnatural round of existence. Others might adjust if a new pattern of regularity were established. But these days of exploration underground lacked any vestige of regularity, a situation that contributed to the overwhelming fatigue and frustration that eventually overcame even the expedition's strongest cavers.

The members of the Crystal Cave expedition are bound in a brotherhood deeper than that known by other cavers. The grueling hours of supply carrying, truly terrifying as well as exhausting for some, the tense anxiety felt for fellow members in perilous climbing attempts and for those who went hungry below ground in the first days while surface teams were commandeered for fighting raging forest fires, the mysterious, sometimes unexplained happenings in the depths of the cave, and the light-hearted banter, the corny humor, and the rollicking cave ballad composed and sung underground—all are precious experiences that cannot be fully shared with others.

The fact that we know so much about them is due to the explicit but succinct notes in faithfully kept logs and to actual, on-the-spot recordings of telephone conversations, routine reports, and radio and television broadcasts, many of them from the depths of the cave. Without these detail-rich supplements to the vivid memories of its leaders this account of the expedition never could have been written.

Transparent in this book is the true explorer's respect, almost reverence, for the handiwork of nature. Even in the most remote parts of the cave, delicate and beautiful formations were protected from damage. They were recognized as being just as precious as the irreplaceable mineralogical and geological marvels in the tourist portions of the cave.

And just as a host of sportsmen above ground have learned the satisfaction of collecting their trophies with a camera, cave explorers bring back glittering formations and fascinating cave animals *on film*. The real objects remain in the cave, *where they belong,* as pristine as when the first human eyes encountered them.

It is impossible to read these pages without recognizing the insatiable drive that spurs man onward, or realizing how frustrating it is to turn back when you can see with the explorer's eyes that there are caves beyond. But never completely solved logistics problems, the growing fatigue that becomes overwhelming—that invisible but insurmountable endurance barrier —and the paralyzing uncertainty as to where you are in the uncharted maze, combine to make the caves beyond unconquerable.

But there will be new assaults on this and other unexplored caves. And if the efforts of the future are carried on with organization as good, with personnel as skilled and devoted, and with a respect for caves and their contents as sincere, the caves beyond offer man an endless but deeply satisfying challenge.

CHARLES E. MOHR
*Past President, The National
Speleological Society*

ACKNOWLEDGMENTS

THE FLOYD COLLINS' CRYSTAL CAVE EXPLORATION was a team effort in which The National Speleological Society was assisted by the fifteen American firms listed below. These companies generously gave both funds and products to make possible the additions to speleological knowledge which the expedition achieved:

Chas. Pfizer & Company, Inc.
Columbian Rope Company
Ditto, Inc.
E. D. Bullard Company
General Foods Corporation
Habitant Soup Company
Linde Air Products Company, a Division of Union Carbide and Carbon Corporation
M & R Dietetic Laboratories, Inc.
Mine Safety Appliances Company
Minnesota Mining and Manufacturing Company
National Broadcasting Company, Inc.
Pillsbury-Ballard Division of Pillsbury Mills, Inc.
Sylvania Electric Products, Inc.
Taylor Instrument Companies
The Coleman Company, Inc.

AUTHORS' NOTE

ADVENTURE BOOKS AND TALES of daring continue to appear on the literary
scene at frequent intervals. We believe that probably the most characteristic quality in any of them is that they are written some months after
the facts, and give a reconstruction of the action as closely as the author
is able to make it.

Facing this at the beginning of The National Speleological Society's
expedition into Floyd Collins Crystal Cave, we made plans for preserving
by tape recording many of the more important telephone conversations
between parties working in the cave and those manning topside posts. Also
recorded were interviews with explorers who had just emerged from the
cave, and reports of the working press who covered the story for their
readers. By using this material as it was recorded we hope to give the reader
a clearer insight into the successes, problems, and failures of those people
who spent a week underground.

Climbing a mountain is a vastly different operation from exploring a cave.
A mountain is there. You can see its summit, which becomes the goal of an
expedition. You can't see a cave's goal, whatever it may be, for as you stand
on the surface of the land, often the only evidence of a cave is a hole leading downward into the unknown. *The Caves Beyond* is the story of the
exploration of just such a hole, and how the explorers pushed the limits of
their knowledge to the limits of physical endurance.

The story is told from two viewpoints by two people who experienced it
from differing positions. Joe Lawrence, Jr., saw it as a leader of the expedition. He saw his years of dreaming and planning being put into effect.
Roger Brucker saw it as an explorer and surveyor. For him it was a challenge hurled up by the unknown.

Some of the raw data of the expedition have not yet been fully analyzed
and evaluated—nor, perhaps, will they be for some years to come, since the
cave explorer in America is forced to pursue his hobby in his spare time and
at his own expense.

To prevent becoming over technical throughout the narrative of the expedition, we have included official reports and other findings in the form
of appendices at the back of the book. While some of the facts have been
entered in the running account as they were discovered, the authors feel
that the reader may wish to examine the official reports in their entirety.

As you read this account, other parties will be penetrating into the remote reaches of Crystal Cave to assault the endurance barrier. They will undoubtedly discover many new waterfalls, rivers, and galleries, but all of them will rely heavily on the findings of the 1954 expedition into the cave. The National Speleological Society is an organization devoted to the study of caves, and the study of Crystal Cave will not be allowed to lie dormant as long as one passage remains unsurveyed and one enigma remains unsolved. This is the determination of the cave explorer. This is the determination of those who have written this account.

JOE LAWRENCE, JR.

ROGER W. BRUCKER

February, 1955
New York

THE CAVES BEYOND

A National Speleological Society explorer rappels into a pit in Floyd Collins' Crystal Cave. *Photo by Robert Halmi.*

Part I

PREPARING FOR THE ASSAULT

by Joe Lawrence, Jr.

FROM DARKNESS INTO DARKNESS

December, 1951

I HAD LOOKED FORWARD for three days to the moment I would see daylight again. Sixty hours can seem like a year when you spend it in the unchanging darkness of a cave. My eyelids were beginning to droop. I had to think deliberately to put one foot in front of the other as I trudged upward toward the cave entrance. The weight of my sleeping bag sank uncomfortably into my shoulder, but fatigue was a much heavier burden. Now I was less than a hundred yards from the entrance, so I moved on, almost automatically, longing to see the sunlight.

For three days I had traveled through the limestone galleries and canyons of Floyd Collins' Crystal Cave, seeing more twisted passages and unexplored chambers than I had ever seen in any other cave. Ahead of me Bill Austin pushed open the heavy wooden door at the entrance. A blast of winter air struck me as I followed him up the steps. There was no sun! I looked up to see if the roof of the cave was still overhead and saw, instead, a million stars. I had completely lost my sense of time. In the timelessness below, the constant dripping of water, the perpetual darkness, and the everlasting silence, all remained unchanged from day to day, from eon to eon.

Jim Dyer met us with a wide grin as we climbed the hill toward his cave office. He was clean-shaven and cleanly clad in contrast to our mud-coated beards and shredded clothing. Jim was in his forties, a slightly bald man with an almost impish smile. Nearly everyone who had come in contact with him talked of him as though he were an old friend, even if the meeting had been a brief one. I looked at Jim through bloodshot eyes and mumbled, "You've sure got a lot of water-cut passage!" Then I flopped to the ground beside my red jeep.

3

The eight of us who had spent three days below were veteran cave explorers, but we had only glimpsed the vastness of Crystal Cave. Its tubes twisting, turning, and intersecting seemed to have no end. Three days in Crystal Cave was long enough to confuse but not long enough to help us comprehend. How big was the cave and where was the end?

I had first heard of Floyd Collins' Crystal Cave when Roy Charlton told me of a trip he had made to the lower levels of the cave in the summer of 1951. Jim Dyer, the manager of the commercial part of the cave, and Luther Miller had guided Roy on that trip. When Roy told me of the passages that seemed to have no end I knew he had seen something big, for Roy is a sharp observer and a good judge of caves. Roy and I had explored many Virginia caves together, and I considered him one of the best explorers I knew. There was no restraining him when he entered an unexplored passage; he would push forward doggedly to the end.

Roy and I had decided to spend three days in Crystal Cave during our Christmas vacation. With three other Virginia cave explorers, we piled into my red jeep and bounced off over the rugged country that separated us from central Kentucky.

Floyd Collins' Crystal Cave lies under an island of private property surrounded by Mammoth Cave National Park. The entire region is a limestone plateau shot through with the longest cave systems in the world, and the plateau itself is dissected by solution valleys dividing the area into ridges, with each ridge containing its own cave system. The Mammoth Cave Ridge is the best known of these, extending for about three miles north and south and averaging three quarters of a mile in width. Directly east is Flint Ridge, about four miles wide and from two and a half to three miles long. Under Flint Ridge lies Crystal Cave along with Salts Cave, Colossal Cave, Great Onyx Cave and several smaller caves, all of more recent discovery than Mammoth Cave. For this reason, systematic exploration of the Flint Ridge system had never been pushed so completely as had that of Mammoth Cave, but the potential of the Flint Ridge system appeared to be far greater in terms of sheer area of coverage.

The jeep rolled through Cave City, passed Mammoth Cave, and followed a twisting gravel road past weathered limestone outcrops that stood above the brown sod. The road led out of a valley and onto the wooded crest of Flint Ridge. Road cuts revealed red earth, characteristic of the cave areas in Kentucky and Virginia. Here and there were outcrops of the red Cypress Sandstone that capped Flint Ridge and overlay the thick beds of cave-bearing limestone.

When we turned into the road leading to Crystal Cave we saw sinkholes extending downward for a hundred feet or more. To the untrained eye they looked like valleys of the type found in many another locality, but these had no streams flowing out of them. Instead, surface water ran into the sink-holes and vanished into the honeycombed limestone where it eventually reached underground streams. The geologists have named this kind of terrain *karst* after the Karst area in Yugoslavia where subterranean drainage in limestone was first studied by European scientists. Since the late 1860's American geologists have studied this classic American example of karst in Kentucky and some of the most brilliant observations on cave development have come from their studies.

The red jeep squealed to a halt in front of one of the three frame buildings at Crystal Cave. A one-story, green-roofed frame building was Jim Dyer's home and office. Dyer and Bill Austin stepped out of the office to see what had arrived. Bill Austin was thin, wiry and wore glasses; not exactly the build you would expect of a cave explorer. But if he was as good as Roy had told me, he was one of the best in the country.

The pair stared in amazement as five men, duffel bags, a tent, a carton of groceries, sleeping bags, rope, a shovel, a camp stove, a rope ladder, flash-lights, and other odds and ends poured from the jeep. Roy introduced us.

"You're early; we weren't expecting you until tomorrow," said Jim.

"Yes, I know. We made better time than we expected," I replied. "The jeep only broke down once."

"I don't blame it," said Austin who was still staring incredulously at the pile of equipment we had unloaded.

"Bill and I won't be ready to take you to the lower levels until tomorrow morning," said Dyer. "But we'll show you the tourist routes just as soon as you have eaten and rested."

An hour later the five explorers from Virginia followed Jim and Bill down a stone stairway into a small sinkhole that was the cave entrance. As we passed through the doorway at the bottom of the sinkhole, Jim flipped a light switch, revealing a passage fifteen feet wide stretching ahead of us. Floyd Collins, who had discovered the cave in 1917 when he crawled into a small hole in the bottom of the sinkhole to retrieve a trap, had dug out the floor so that tourists could walk in and view the cavern.

We followed our guides along the passage that was smoothly roofed by a perfectly flat layer of limestone. This was immediately below the layer of Cypress Sandstone. The walls pinched in, then retreated as we passed a sign warning visitors not to touch the formations. We passed rocks piled up at the sides of the trail by Floyd and his brothers during the 1920's.

We came to the brink of Grand Canyon where our flashlights barely pricked the darkness. Then Jim switched on floodlights, revealing a vast,

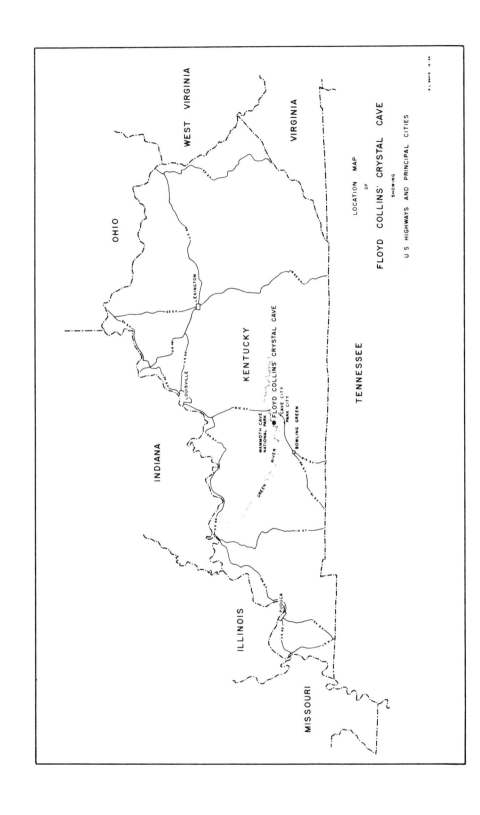

LOCATION MAP

OF

FLOYD COLLINS' CRYSTAL CAVE

SHOWING

U S HIGHWAYS AND PRINCIPAL CITIES

crescent-shaped chamber that extended out of sight around the bend. From our vantage point near the ceiling seventy-five feet above the floor, a trail snaked downward between walls thirty feet apart. This width increased to almost a hundred feet in the center of the room. The dull gray of the passage behind us was replaced by golden brown walls.

At the bottom of Grand Canyon the room looked even larger. We could see a large passage extending off at right angles to the big room in both directions. In this cross-like intersection was Floyd Collins' tomb. A bronze casket rested on the floor in front of a granite marker which bore Collins' epitaph. It had been his wish that he be buried here in the cave he loved so dearly.

Floyd Collins was quick to realize the commercial value of the cave he discovered. He dug out the entrance tunnel and constructed paths through the large passages so that he would be able to show the cave to tourists. This inevitably plunged the Collins family into the keen competition that exists between the many commercial cave operators of the area, men who were secretive about what they had and suspicious of their neighbors. It is said that Collins found passages he thought led to the property of others. He feared that another part of Crystal Cave might be shown to the public by a competitor using another entrance, or that vandals might slip in the back way to destroy the beautiful travertine and gypsum formations that attracted people to his cave. There are rumors that he blocked with boulders the passages he thought were a potential danger. We can only speculate on whether this is true, for Floyd was careful what he told others about his cave and didn't always confide even in his brothers. Many secrets died with him.

Floyd Collins was first and foremost an explorer. At every opportunity he wormed his way through the crawlways beyond the tourist trails to explore his limestone labyrinth. He worked alone, partly because secrecy was necessary and partly because that was the make-up of the man. With a kerosene lantern and meager rations he would disappear for days at a time to push into the unknown, apparently with little concern for the hardships he faced or even for his own safety.

Floyd was driven by the cave explorer's insatiable curiosity to see what lay around the next bend in the passage ahead. But he also realized the commercial value of another entrance closer to a highway, if he could control that entrance. To reach Crystal Cave tourists had to drive several miles from the main road past compelling signs advertising the caves of competitors. Floyd thought he could alleviate this disadvantage, for he suspected

Sand Cave, about five miles to the south, connected with his Crystal, and Sand Cave was close to a main highway. In the crawlways of Crystal Cave he squirmed toward Sand Cave, but so far as we know did not find a connection. Then he wriggled through Sand Cave, seeking to make the connection from that end. He did not succeed, so he tried again.

On Saturday, January 30, 1925, the world read the name of Floyd Collins as newspapers carried the story that he was trapped in little-known Sand Cave in the hills of central Kentucky. The reading public seemed to hold its breath as it waited to hear of the rescue of this spirited explorer. In Louisville, Kentucky, William Burke ("Skeets") Miller, a young cub reporter for the *Courier-Journal,* requested to be assigned to the story. After some persuasion his city editor agreed, but warned, "Now don't you go crawling around in any cave."

Miller arrived at the cave on Monday, February 1, as a cold gray dawn was breaking. There were only three local people standing near the bleak cave entrance. When Miller quizzed them about the tragedy, he was told, "There's the cave. Go down and get your own story." He took up this challenge and slipped his small, hundred-and-ten-pound frame into the crawlway under jumbled boulders.

With flashlight in hand he squirmed forward on his stomach like a crawfish. As he crawled through the damp tunnel that appeared ready to collapse, he came to a short drop. He slipped and fell headfirst into the hole, landing on a soft, wet mass. The mass groaned. It was Collins! Miller squirmed to the side and asked the trapped man how he was, but Collins jabbered incoherently. Miller slipped one leg in beside Collins but could go no farther. The crawlway was so small that Collins' body blocked it. A boulder Floyd had dislodged rested on Floyd's leg, holding him prisoner in this dark, dripping place. Collins lay on his left side, his head toward the entrance, so that, although rescuers could not squeeze past him to free his leg, they could feed him.

Skeets helped Floyd adjust the burlap that had been placed about the trapped man's head to protect him from the discomforts of dripping water. Then Skeets Miller returned to the surface to file his story.

By Monday afternoon and Tuesday a crowd of "outlanders," as the natives called the reporters and curiosity seekers, had gathered at the cave. The hopeless drama of a man fighting for his life beneath the surface of the earth unfolded against a backdrop of cut-throat newspaper reporting. Through an error, Miller's first story was routed to a rival newspaper and the *Courier-Journal* reporter's account was spread across its front pages. Later, some reporters claimed the cave accident was all a hoax arranged by

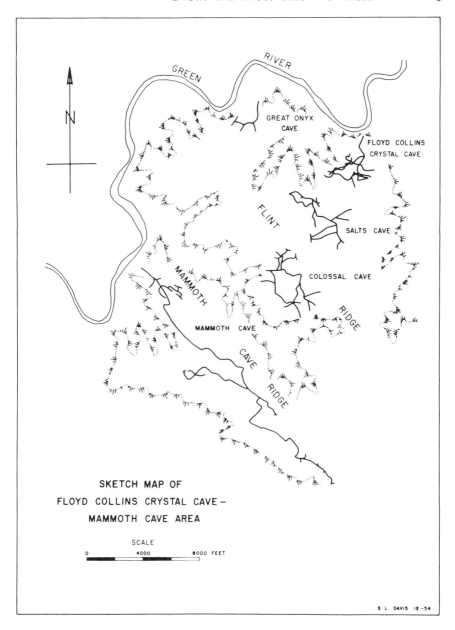

SKETCH MAP OF
FLOYD COLLINS CRYSTAL CAVE —
MAMMOTH CAVE AREA

SCALE

S L DAVIS 12-54

Collins, Miller, and the *Courier-Journal.* A rumor started that each night Floyd Collins left the cave by a secret exit, returning in the morning to feign the accident for its publicity value. Things went so far that film was even switched by an unscrupulous photographer, and Charles A. Lindbergh, who had been sent to the cave to fly pictures to Chicago, unsuspectingly delivered blank film to his paper instead of the photographic record of the tragedy.

As the crowds grew, martial law was declared and a military guard placed at the cave entrance. Miller was named pool correspondent and was the only reporter allowed in the cave.

Skeets Miller was touched by the plight of the doomed man. He crawled under the dangerous sandstone boulders that roofed Sand Cave to bring comfort and to attempt rescue for the helpless Collins. He brought in hot soup that Collins hungrily drank, and he laid a string of electric lights in the cave. He wrapped a bulb in burlap and placed it against Collins' neck to keep him warm. The delirious man understood the kindness and tried to express his appreciation.

Then Miller dragged a jack and a crowbar into the cave to attempt to raise the boulder, but he could not get enough leverage to move it. Next he tried to slip the jack under the boulder. Working patiently in terribly cramped quarters, Skeets finally forced the jack in place. Floyd's plight was not hopeless! The jack would surely raise the boulder. Miller started working the handle up and down; the boulder moved a fraction of an inch. Then the jack jumped free and the boulder settled back into place. Tediously, Miller poked and struggled until he got the jack in place again; but it only slipped away when he started to raise it. Again he tried and again he failed, but he kept at it until he had tried every possibility and had exhausted himself. Reluctantly, Miller returned to the surface.

Shortly after Miller had returned from feeding Collins, part of Sand Cave collapsed. Miller re-entered the cave to determine the seriousness of the rock-fall and found the passage blocked a short distance from where Floyd lay trapped. Miller called out to Floyd. From the other side of the barrier Floyd's answer came back, "Come on down; I'm free."

"If you're free, reach the bottle of milk I've placed in the crevice over your head," shouted Miller.

After a short silence a dull monotone came back through the rock-fall, "No—I'm not free." These were the last words anyone heard from Floyd Collins.

With utmost haste, a shaft was started on the surface over the spot where Floyd lay trapped. Twelve days after Miller had last spoken to Floyd the diggers broke through to find that Floyd Collins had died of exposure. His body was removed and eventually placed in a bronze casket in the center of Grand Canyon Avenue in Floyd Collins Crystal Cave.

I looked up from the casket when Jim Dyer said, "Come on. I'll show you some formations that will knock your eye out."

Jim Dyer and Bill Austin led the way up a trail that climbed out of the far end of Grand Canyon. We rose to the level of the entrance and snaked along a narrow trail past the crude wooden wheelbarrow Floyd had used to haul out rock and earth excavated to make accessible new trails. It was brown and rotting in the damp cave air. Jim explained that we were on the Helictite Route.

Helictites (top) grow on stalactites (hanging downward) and columns (center). On the floor are stalagmites. Constantly dripping water continues to deposit calcium carbonate on live cave formations. *Photo by James Dyer.*

Ten minutes later we stood looking at a dazzling display of stalactites and stalagmites. The white and reddish-brown stalactites hung like icicles from the ceiling. Water dripped slowly from them and splattered on stalagmites that reached upward from the floor. This water had worked its way from the surface through cracks in the limestone and had become saturated with calcium carbonate, the principal constituent of limestone. As each drop hung from the end of a stalactite some of the water evaporated and a minute deposit of calcium carbonate, or travertine was left behind.

More of the water evaporated after the drop hit the stalagmite below and left a slight deposit at the top of the stalagmite. In this way the stalagmites grow upward and the stalactites grow downward. We saw some columns reaching from floor to ceiling that had been formed when stalactites and stalagmites grew together.

Then we saw the helictites on the walls and ceilings alongside the trail. These rare formations were everywhere. They curled upward and downward, stretching their white stone tendrils in every possible direction. Some formations looked like shredded coconut, others like Medusa's snaky locks. Even at that time scientists were unable to explain their growth satisfactorily. Very recently the theory has been advanced that they are formed under dry conditions by the growth of wedge-shaped aragonite crystals along a spiral axis.

We retraced our steps to the bottom of the Grand Canyon and turned left into the side passage Floyd had explored the day he discovered the cave. The passages and galleries had been formed below the water table where water filled every crack in the rock. The water dissolved the limestone, enlarging the cracks into caverns. Caves thus formed are drained when geologic uplift and downcutting streams on the surface cause the water table to drop below the level of the caverns. The air-filled caves are sometimes further enlarged by streams flowing across the passage floors eroding and dissolving more limestone. While cavern building is going on in one part of a cave another part may be filling up. Fine silt car-

Helictite formations line the walls and curl in every direction in one portion of the tourist route. Irregular crystalline growth along a spiral axis has apparently made the fantastic shapes. *Photo by Wm. Austin.*

ried down by water percolating through cracks on its way to the water table may leave fills of clay in submerged caverns. Streams flowing into air-filled caverns above the water table may leave deposits of silt, sand, and gravel.

Up and down we went over tourist trails, then stopped while Jim directed our attention to gypsum "flowers"—crystals of calcium sulphate—growing under a ledge in the Flower Garden. Many of them resembled lilies or asters, extending featherlike petals downward. They were white, or tinged with delicate yellows and tans.

After photographing the gypsum, we moved on over blocks of breakdown, sharp-edged boulders that had dropped from the ceiling of the passage long before man had entered the cave. The passage itself averaged about thirty feet wide and varied in height from ten to thirty feet, depending on how high the piano-sized boulders were piled. Our flashlights wouldn't pierce the blackness ahead.

"This is the Valley of Decision," said Jim. "Floyd came to this spot with a kerosene lantern on the day he first entered. He paused here, just as we are doing, only there was no trail then. His lantern wouldn't begin to light up the area ahead of him, so he had to make a decision, whether to go on, or come back later with more light. He decided to go on, and this is what he found." Then Jim turned a switch and flooded the area with light. The Valley was an impressive room floored with red sand. It was about sixty-five feet from floor to ceiling at its lowest point. The path climbed out of it on the opposite bank over a hundred feet away.

The rooms were bigger now, averaging fifty to sixty feet wide. We came to a high-ceilinged room where a chaotic collection of boulders littered the floor and reached the ceiling at the far end, blocking further progress in the direction we had been traveling. This was called the Devil's Kitchen.

Around the corner we stopped at a trickling stream of water falling from the top of a dome into a shallow pit below. Jim offered us a drink. On we went into a narrower passage, about ten feet wide and twenty feet high. Because the walls were covered with gypsum formations varying from flowers to great sheets covering entire wall areas, Floyd had named it the Gypsum Route. The trail meandered along for nearly a half mile in this manner and at each bend Jim or Bill would point out odd and beautiful formations. One resembled an opossum's tail jutting from the ceiling, another looked like the profile of a cow.

Ahead of us, where the electric lighting ended, a boulder apparently blocked the passage. Jim ducked under it and we followed, coming up on the other side. Jim pointed to a small hole in the floor on our left and said, "You go through there tomorrow. That's Scotchman's Trap."

"How far are we from the entrance now?" I asked.

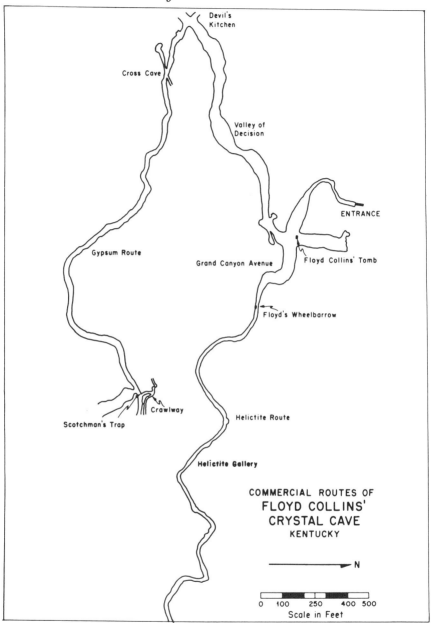

COMMERCIAL ROUTES OF
FLOYD COLLINS'
CRYSTAL CAVE
KENTUCKY

Scale in Feet

"About a mile," replied Jim, "But you haven't seen anything compared to what you'll go over tomorrow to reach the lower levels."

As we returned to the surface we talked of the earlier explorations of Crystal Cave. Harry Dennison and Ewing Hood probed Crystal Cave after

Floyd Collins' death. For years they searched for one of Collins' great dis-
coveries—a room reputed to be a mile long, known as Floyd's Lost Passage.
Finally they rediscovered this vast gallery in 1941. After World War II, Jim
Dyer, Luther Miller, and Bill Austin pushed on into whole systems of passages Floyd Collins had never seen. But the cave seemed to have no end.
Austin and Dyer could stay in the cave only as long as they could go without sleep. Each time they went to the lower levels beyond Scotchman's
Trap they had to turn their backs on unexplored passages when the time
came to return to the surface. It took six hours to reach the limits of Collins'
penetration and six hours to return. This left little time for advancing the
limits of exploration. Jim Dyer was further limited since as manager of the
cave he worked seven days a week taking care of the tourist business and
could usually explore only at night. Always time and energy would run out,
and the explorers would return to the surface more and more bewildered at
the complexity of the cave.

It was with this knowledge that Roy and I had planned our three-day trip.
In most caves we were able to push to the physical limits of the passages,
but in Crystal Cave explorers turned back at their own endurance limit
while the cave went on. Roy and I felt that we could advance this endurance limit and perhaps reach the limits of the cave if we managed to
establish a camp in the lower levels where we could get food and rest before continuing into the unexplored.

In the morning after breakfast we made a final check of the equipment
we had packed the night before; then we descended into the cave. Eight
of us filed along the mile-long tourist trail to Scotchman's Trap. We were
the five Virginia cavers, an Indiana caver who had arrived late the night
before, and the two Crystal Cave veterans—Austin and Dyer.

At Scotchman's Trap we paused to light the carbide lamps on our hats.
Since the string of electric lights ended here, our only light beyond would be
what we carried on our heads. I strapped on knee pads. I had never used
them in a cave before, but I had been advised to wear them, so to save
argument, I did.

Bill Austin slipped through the hole in the floor that led to the Crawlway. Two others followed; then we passed our equipment through. We
were loaded heavily with food, a gasoline stove, carbide, and sleeping bags.
We carried just four sleeping bags, for we had been told the difficulties
of getting supplies to the lower levels were too great to carry any more
than absolutely necessary. By sleeping in shifts we felt we could get along
comfortably.

One by one men and bundles passed through the small hole into the

Crawlway. We stationed ourselves about four feet apart and moved the equipment from man to man over a thirty-foot stretch of Crawlway. Then we dropped into a narrow passage with a high ceiling and walls fourteen inches apart. Ragged rock protrusions clutched at our clothes and we pulled our equipment along. We began to sweat even though the temperature was a cool 55 degrees.

Now we were on our hands and knees, then on our stomachs as I dug my elbows and knees into the sand floor and squirmed. Dust crept up my nostrils. The sleeping bag I was pushing before me snagged on a projection, but ahead of me Roy Charlton saw my plight and turned to free the bag for me. Sweat streamed down my face, spattering on my gloves. Caving in Virginia had never been like this.

I now found out why the knee pads are necessary. Already my knees throbbed, and we had come only about three hundred feet. On we went through the S Curve where we had to twist our bodies in two directions and through the Keyhole that just barely allowed us to pass.

After several hours of crawling and squirming we came to larger passages with canyons in the floor that we had to straddle. We came to climbs up and climbs down. We clambered over boulders, then we crawled again.

When we entered a low wide room Bill Austin said, "Be careful, you are on top of X Pit." At one end of the room there was an eight-foot drop into a lower passage, and at the bottom of this a hole went down at an angle of 45 degrees. In ten feet it opened into the void of X Pit. We moved into the lower passage while Bill stayed eight feet above us to pass down the gear. Roy caught what he thought was the last sleeping bag, then turned away just as Bill threw down another. It hit with a thud and rolled into the 45-degree-hole toward X Pit. I watched in horror as Austin leaped down the eight-foot drop in an effort to catch the bag. He straddled it momentarily as both he and the bag skidded down toward the 140-foot shaft. Austin stopped. The bag continued into the blackness of the pit.

"Joe," said Austin calmly, "you and the others stay here. Roy and I will try to fish the sleeping bag out." They disappeared into a crawlway leading in corkscrew fashion to the bottom. In a half hour they were back with a torn sleeping bag.

Beyond X Pit we encountered more climbing, more canyon straddling, and the inevitable crawling. Dyer was leading now as we dropped into a muddy corridor that he told us was Mud Avenue. When we stopped to rest I noticed that Austin and Indiana were no longer with us, but I was too tired to wonder what had happened to them. "Let's go, boys. Time's a wastin'," said Jim, and we were on the move again.

I had completely lost my sense of direction. We had passed so many intersections, made so many turns, gone up and down so many times. I wondered if I could find my way out of this three-dimensional maze. I had never been lost in a cave, and I had been in some confusing ones before. It probably wouldn't be easy, but I thought I could find my way out if I had to. I was glad Dyer was along, though.

We climbed up out of Mud Avenue and suddenly we were upon the vast expanse of Floyd's Lost Passage. After eight hours of traveling through cramped crawlways and narrow canyons, a wide, smooth-floored passage in which we could stand was especially welcome. This "walking cave" looked so good to me that I ran up and down in spite of my weariness, then sprawled out on the sand floor, out of breath.

"I smell food cooking!" said Roy.

"Not down in this hole," I said gloomily. "You must be delirious." But then I smelled it. It was beef stew!

At the Valley of Decision Floyd Collins chose to continue his exploration in spite of his feeble lantern. *Photo by James Dyer.*

We were on our feet and moving fast toward the aroma, for we had eaten no more than a candy bar apiece in the past eight hours. Jim grinned broadly, but said nothing.

We rounded a bend in Floyd's Lost Passage and there before us was a glowing underground camp. Canned food was stacked neatly on a rock shelf. A gasoline stove burned merrily under a pot of stew. Candles spread soft, yellow light and cheer over the scene. Austin bent over the stew pot with a spoon and Indiana sat nearby drinking a cup of coffee.

"Crystal Cave is sure full of surprises!" muttered Roy.

"But Austin is fuller," I added.

Austin, Dyer, and Indiana laughed heartily at the amazement of the Virginians. Then all of us ravenously ate the hot, nourishing food.

After supper I crept into a sack as several others bedded down for the night. Half of the crew stayed up to spend the night exploring, the price they paid for our lack of sleeping bags.

In the morning when the night crew crawled into the sleeping bags, Jim Dyer took me and two others out to explore leads off of Mud Avenue. We explored several that either led to intersections with other unexplored leads or doubled back into Mud Avenue. But there were many more leads we did not have time to investigate. After lunch Jim and Indiana left the cave. The tourist business required Jim Dyer's presence on the surface.

On the evening of the second day Roy Charlton and Bill Austin went with a party toward Bogardus Waterfall and the endurance limit we hoped to puncture. In this area they entered a low crawlway, virgin cave. They went forward on hands and knees for five hours, then they reached that invisible barrier where the passage continued onward but where they couldn't. Fatigue forced them back. Wearily, they returned to camp just as Jim Gosney and I were arising from a good night's sleep.

Time was running out, so Gosney and I decided to start out of the cave with some of the equipment that had to come out. The others would have to get some sleep before they would be ready to travel.

"You won't be able to find your way out," warned Austin who sat on a sleeping bag unlacing a boot. "Too many passages between here and the entrance."

"Maybe," I said, "But I've been in a lot of caves and haven't seen one yet that could lose me."

"We'll be along in about six hours to lead you out," said Austin.

"We'll be on the surface by then," I said. "But to play safe, we'll leave notes along the way so you'll know the route we traveled as you come along behind." Then Gosney and I walked off toward Mud Avenue.

When we came to the junction where we left Mud Avenue I wrote a note. "We are turning off here. We will leave another note at Floyd's Jump-Off." Then at Floyd's Jump-Off I tore another sheet from my muddy notebook and wrote: "We are finding our way all right. We will leave a note at the X Pit." We passed X Pit, leaving an optimistic note, and headed on toward Ebb and Flow Falls. There I left a note saying we were on our way to the Keyhole. Thirty feet on we came to a confusing junction of passages. Nothing looked familiar! Systematically, we tried each passage but each led to another junction, none of which we recognized.

Back at Ebb and Flow Falls I added to the note: "We are having trouble finding our way. Will try again and return at 3:00 P.M." We did try again, and this time found more confusing passages and junctions. Again we retreated to Ebb and Flow Falls. We added to the note: "We are lost. DON'T leave without us. Back at 4:00 P.M."

At four we returned to the Falls. We worried; perhaps this was not Ebb and Flow Falls after all! We sat down to wait, speculating on how long our supply of carbide would last.

When Austin, Charlton and the others arrived, we breathed more easily. It was a pleasure to crawl in Austin's dust the rest of the way out.

When I reached the surface and lay down beside my red jeep on that cold December night, I tried vainly to fit together the pieces of the cave I had seen. I wanted desperately to understand Crystal Cave, but I could not.

Jim Dyer looked at me as I lay there exhausted but wide awake, and with a playful smile he asked, "Well, Joe, do you want to try it again?"

"Let me catch my breath for a couple of days," I said, "then I might do it. But if I ever go back in there I'm going to stay a week, and I'll have enough people along to really explore that hole."

The smile left Jim's face and he looked at me intently, for he had not expected such an answer from one who had just spent three days in the arduous passages of Crystal Cave. My answer also reminded Jim of his own insatiable desire to understand the cave. He knew infinitely more about it than I, but his knowledge left him with more questions about the parts of the cave he did not comprehend.

Bill Austin looked at me and said, "When you are ready, Joe, we're with you. There's a lot down there we don't understand."

My experience in Crystal Cave had started a fire within me, a fire I shared with the veterans of Crystal. Others had come before and others would come after, and their taste of Crystal would start fires within them.

Bob Handley, an ardent explorer who had worked with me to conquer the lower levels of Starnes Cave in Virginia, had seen Crystal Cave and felt the urge to penetrate the veiled unknown of the Crystal Cave system. Three cavers from Ohio, Roger Brucker, Phil Smith, and Roger McClure, were to view the lower levels of Crystal Cave for a day and leave with the burning desire to return and to understand. It was in the minds of those who had seen the cave and been turned back by the endurance limit that the 1954 Floyd Collins' Crystal Cave Expedition was born.

2

RUN LIKE HELL

WHEN JIM DYER RESIGNED as manager of Floyd Collins' Crystal Cave in 1952, Bill Austin took the job. He moved into the house a hundred yards from the cave entrance beside the old Collins homestead, which was a square frame building like hundreds of other unpretentious Kentucky homes built early in the century. It was here at Crystal that Bill had cut his teeth on cave exploring under the tutorship of Jim and a handful of others. By this time, he had spent hundreds of hours in Crystal Cave's depths, and with each trip to the lower levels he brought out penetrating questions that no one could answer.

With the challenge right under his living room he couldn't forget it, even though the tourist business made great demands on his time and energy. Where did B Trail go? What was at the end of Mud Avenue? Was there unexplored cave at the top of Fools Dome, an intriguing alcove within easy reach of Floyd's Lost Passage, but a target for skilled rock climbers to reach? On many occasions he had stood at the bottom of it looking up into the small hole leading to the unknown, wishing that his occupation was cave exploring and that he could devote full time to alleviating his "cave itch."

The drive to find out what was there took him below again and again, but he realized that he could not explore this cave to its limits with just two or three companions, no matter how many times they returned. He began to turn over in his mind the idea of a week-long exploration trip, the same idea I had expressed when our party emerged baffled and unsatisfied after three days in the black wilderness of intersecting passages. The network didn't spread in just one direction; it went six: north, south, east, west, up and down. Each passage led to more passages and each canyon led to more canyons. It seemed to him, as it had to his predecessors, that discovery was a logarithmic progression in which you squared the unknown each time you made a discovery.

There are more crawlways and passages than explorers have time or stamina to explore. This is Crystal Cave's endurance barrier. *Photo by James Dyer.*

Bill lay awake many nights over this mystifying grid-work. It would take a large sustained expedition to wrench free the secrets of Crystal Cave. Perhaps then he would find out where the canyons went and where the streams emptied. He might find the fabulous Big Room two earlier explorers had babbled about. They had described an immense chamber so large that a two-cell flashlight failed to reach the walls when they had stood in the center of it. In the middle, they said, was a large boulder, so big that they got lost when they started walking around it. Was this a Paul Bunyan story, a creation of their minds? Or did it really exist, waiting only for some explorer to enter the right crawlway?

And what about the Spaniard's story? "The Spaniard," now a legendary figure but well-remembered by Jim Dyer, had started around the world in quest of adventure. When he reached Kentucky he sought Crystal Cave and wangled permission to explore the lower levels. He went in with blue chalk in his pockets to mark the way and a blanket roll over his shoulder. He traveled through the Crawlway, leaving a trail of prominent blue arrows behind that can still be seen. Three days later he emerged, raving about the vistas he had seen. He had seen roaring torrents of water. At one place he had tied his blanket to a projection of rock in the ceiling and had swung across the raging river on the blanket. Then he had found majestic drapery formations encrusted with crystals of selenite. Finally, he had been turned back at the shore of a deep lake too wide for him to cross, but on its opposite banks he had seen travertine columns eight feet high and a glistening city of stalactites and stalagmites.

Austin had never seen such bodies of water in Crystal Cave. Neither had Jim Dyer. Generally speaking, the known portions of Crystal Cave were too dry for the building of large travertine formations. Was the Spaniard's tale fact or fiction? Until Austin had satisfied himself that he had seen *all* the cave, he couldn't be sure.

Natives in the area told stories too, describing how Floyd Collins would disappear into the cave for days at a time, then turn up in some farmer's cornfield miles away. Floyd would stagger to the nearest farmhouse and ask where he was. How much was legend? Could an expedition find other entrances to the system? Bill Austin had to know.

Dr. E. Robert Pohl, a geologist and the son-in-law of the cave owner, was just as eager to have a well-organized expedition penetrate the lower levels. They both knew that if there was any group that could handle such an expedition it was The National Speleological Society. Both were active in the society, as are most serious cave explorers around the country.

It was just before the society's Louisville Convention in April, 1953, that N.S.S. President Charles Mohr fell completely under the spell of Crystal. For several evenings from dusk to midnight he set up his batteries of floodlights along the formation-rich passageways and filmed the mineralogical wonders. Then, transferring activities to an all-night restaurant at the bus depot in Cave City, he would sit with Bill Austin and Jim Dyer almost till dawn, listening to their descriptions of Floyd's Lost Passage, and other faraway places in the lower levels.

In their enthusiasm, Bill, Jim, and Doctor Pohl invited Charles to get a group of the best cavers together immediately and make a descent into Scotchman's Trap. "You've got to see it for yourself," they said. But the formal meetings of the Convention would be starting within two days, in Louisville. After the meetings the cavers would be heading back home.

Furthermore, such a venture would require much time for planning. So Charles Mohr had to pass up the chance.

But he believed there must be a concerted effort by an expedition of competent cave explorers. Certainly if the N.S.S. were to go in and look, they would look long and deep. When they did th_y wouldn't add to the store of legend; they would scrutinize, measure, photograph, and record what they saw. They wouldn't describe raging rivers, but a stream of twenty-five gallons per minute. They wouldn't rave about vast, vague rooms; they would record a chamber as 5000 feet long. And they wouldn't bring out tales of mysterious subterranean creatures; they would report in detail on the habits of *Pseudanophthalmus tellkampfi* in Crystal Cave environs. Dr. Pohl and Bill Austin agreed to invite formally the N.S.S. to make a full-scale expedition into Crystal Cave, and in September, 1953, Bill wrote the letter to Charles Mohr.

Mohr conveyed the invitation to the Board of Governors of The National Speleological Society. After a lively discussion of the problems posed by the cave, the board appointed a committee consisting of the president, Charles Mohr, a past president, Bill Stephenson, and the vice president in charge of organization, Joe Lawrence, Jr. They were empowered to investigate the possibility of staging an expedition, and if they thought the idea feasible, they were to undertake the venture as soon as possible.

Charles Mohr cornered me after the society's board meeting and we decided to get together two days later in Philadelphia. He wanted to hear all about my previous three-day trip to the lower levels. At this meeting, Mohr read over my report of our last trial by exhaustion.

"To push beyond the present endurance limit," I explained, "we *must* establish a camp deep inside the cave. The explorers will have to push beyond that camp, using it as a base of operation for food and sleep. Support teams will have to be based on the surface, so they can carry on a continuing resupply of the camp."

"You tried something like that in 1951, didn't you, Joe?" said Mohr.

"Not exactly. We carried all our supplies with us when we entered. Not enough to last more than three days, so we had to come out before we were able to take full advantage of our underground camp."

Mohr screwed up his face. "Then, how many days do you think we ought to plan on staying down there?"

"I know there's enough cave to keep us busy for a week, maybe more."

"Then we ought to plan on at least a week underground," said Mohr. "And it sounds to me as though we're going to need twenty, maybe thirty people to do it. They'll have to be experienced, of course. We need the cream of the society. Our scientific people will jump at the chance to study the geology and fauna."

"We'll need surveyors, a lot of them. You know, Charles, exploration doesn't mean anything unless we survey our discoveries and know exactly where they are."

"Communications, Joe. Could we use telephones?"

"I don't know. I don't see how we could get wire through the Crawlway. It's a struggle just to get yourself through. But we have to have communications on an operation this size . . . wires will have to go through the Crawlway no matter how difficult it is."

The Collins home was like a hundred others in this desolate region of Kentucky in 1917. Floyd Collins discovered his cave in a sink-hole three hundred feet from his front porch. *Photo by James Dyer.*

Mohr got up from his chair and began to pace the room. "If someone gets hurt . . . we have to have a doctor. More than a doctor; he's got to be much more than that. He has to be a caver too. Just any doctor couldn't be counted on to get a victim out through the Crawlway. Probably would become a casualty himself."

I mentally ran down the list of doctor-cavers I knew. "I can think of two who'd fill the bill."

"We'll get one," said Mohr. "But most important, we need a leader, and soon."

"I know of only three men who could do the job," I said. "They all have the ability to lead and plan the thing, and they know what they're up against in Crystal." Then my objective viewpoint crumbled into a heap of scattered thought. I silently wished I could lead this expedition that would probably be the largest of its kind ever attempted in America. We would do things that had never been done before. We would overcome obstacles that loomed impossible now. But most important to me, we would feel the thrill of making the first set of footprints across the smooth dirt floor of passages deep within the earth. To find out where the lower levels lay would be the fulfilment of a dream born three years ago on a cold winter night at the mouth of Floyd Collins' Crystal Cave. Mohr was talking now.

". . . how about it? Will you lead, Joe?"

I turned cold and tingly all over. "Why . . . ah . . . Charles, I'd like nothing better in the world. You know how I feel about it. Yes, I'll lead."

Charles looked across the room at my wife who had been sitting silently, listening. For two hours I had forgotten she was there. "Frankie, what do you think about all this?" asked Charles. I somehow hadn't told her the purpose of the meeting tonight.

With qualms I looked toward her. "I'm afraid I haven't told her about all this. Frankie, what do you think? Would you be unhappy if I spent a week underground?"

With an understanding smile she said, "Joe, you know I wouldn't object. You were caving before you married me, and you probably will go on caving until the day you die. You'd better go on this trip." I felt much better.

I started planning immediately. In the conversation with Charles Mohr we had sketched out the over-all objectives, but there were many details to work out. I thought there would be hundreds of them. I was wrong. There were thousands.

I began contacting key personnel in the society. There were experts in all phases of speleology who would have to be interviewed. I wrote to Bill Austin with the wonderful news and was happy to get back an enthusiastic letter. Some words at the bottom of the page loomed large in my mind, for they threw up the biggest challenge of all:

"We have given the ball to The National Speleological Society. It's up to you boys to run like hell."

Could we do it? Could we put thirty people hundreds of feet underground in an unexplored cave and not lose any of them in the labyrinth? Could we do it without risking tragic consequences? French explorer Marcel Loubens had met his death in a cave in the Pyrenees because of faulty equipment. Floyd Collins died an agonizing death trapped in Sand Cave because he didn't follow the rules. Could a large party follow the rules? Or would we fail because somebody "dropped the ball?"

I made up my mind to run like hell.

3

BLUEPRINT FOR ACTION

S INCE AN EXPEDITION of the type we proposed had never, to our knowledge, been undertaken before, I reviewed the literature on the subject of expeditions. A study of European literature on speleology, especially reports of cave exploring in England and France, revealed to me that their problems were always different from ours because of the nature of their caves. Often these explorers were confronted by the problem of "vertical caving," descending great depths through a shaft-like well to reach the cave where exploration would begin. More often than otherwise, what lay beyond the surmounting of the entrance pitch was a cave comparable in proportions to the majority of American caves. For this reason, European speleological efforts had been directed toward quick and efficient methods of getting to the bottom and getting down to business.

In reading accounts of some of the activities of the seventy or more autonomous cave exploring societies in Britain, time and again I discovered three-, four-, and even five-day trips to "potholes," in which one or two days were devoted to the physical problem of rigging the entrance. Elaborate plans were laid involving the use of hundreds of feet of cable ladders, block and tackles, winches, and similar deep-cave equipment. In some cases, twenty or more people would turn out and would be pressed into service in ways unthinkable to us. Men would be stationed on narrow ledges at the top of ladder-rigged drops, sometimes under soaking waterfalls. Their job was to belay safety lines for the relatively few chosen to penetrate the horizontal part of the cave. Eight hours of waiting on such a ledge often brought men to the endurance limit, but more often than not, it brought the explorers working below to the end of the cave. All hands would then turn to for fellowship around a roaring campfire.

French cave explorers also engaged in this activity on a grand scale in the Pyrenees near the border of Spain. Sheer-walled shafts, or *dolines,* in that great limestone massif intrigued them, challenging them to go deeper and deeper. Toward this end, French planning had been directed, in the main, toward perfecting intricate winch mechanisms which somehow seldom lived up to their designer's expectations. To be sure, both British and French cave explorers were skilled in climbing techniques, more so than the majority of American cave explorers. But that was not the problem confronting us.

The closest parallel to our operation was found in reports of the French attempt to explore Padirac, a French commercial cave containing a river of the same name. By elevator the explorers descended to the river level, then launched their rubber boats into the stream. Tents, sleeping bags, stoves, and other equipment became drenched when the boats were hauled over the succession of natural dams blocking navigation. At one endurance limit the intrepid explorers pitched their tents on a breakdown bridge in the main passage. From this base of operations they launched daily assaults upstream in an effort to reach the uttermost limits of the river. Always they were turned back by fatigue and lack of supplies.

The problem of Padirac was clearly not the problem of Floyd Collins' Crystal Cave. In the first place, heavy equipment such as tents and large, two-man rubber boats could not be carried into the depths of Crystal. The Crawlway, a natural barrier a quarter of a mile long, dictated that. In Crystal, there was not the convenient goal of penetrating one long passage. Instead, there were scores of interconnected passages all of which had to be explored in spite of their confusing complexity and man-killing length.

In the literature of mountain climbing we found valuable information on logistics which would stand us in good stead. One answer seemed to lie in predicating our plans on a series of two or more camps within the cave, depending on the results of reconnaissance, space for such camps, and the presence of an adequate uncontaminated water supply. Unfortunately, we could not draw on native Sherpas and porters to do the back-breaking work of moving supplies. We would have to travel light, perhaps using the best explorers we had to insure even a steady trickle of supplies to those working below.

I could not trust my own judgment alone to plan the details for the undertaking, so I sought expert assistance from specialists within The National Speleological Society. I asked M. Girard Bloch, one of our foremost rock-climbing authorities and students of equipment, and John D. Parker, a speleologist and safety expert, to help me in preliminary planning. Both men were eminently qualified in their respective fields, so I was more than happy when they agreed to meet with me in Philadelphia.

At these planning meetings we listed the objectives of the Crystal Cave expedition in clear-cut terms. The over-all purpose would be to study the Floyd Collins' Crystal Cave system in all its aspects. This meant studying geology and hydrology of the cave in particular, and describing the plant and animal life found in it. We would study cave "weather," meteorological aspects of air pressure, temperature and air movement. We would attempt to determine the extent of the cave and the relationship of the passages to each other by surveying as much as possible in the time allotted. This alone would carry us into the unknown beyond the endurance barrier. We wanted the expedition physician to study the effects on individuals of a prolonged stay underground. His findings would be of significance to future expeditions of this type. We realized that as we went along we would resolve many problems which would go on record to aid planning staffs of future ventures.

With a lucid statement of our objectives on paper, we were able to settle down to the job of determining our requirements in personnel, time, money, and equipment. My earlier discussion with N.S.S. President Charles Mohr had already indicated a possible complement of from twenty to thirty persons. Now, it seemed that thirty would be the absolute minimum for consideration, and that to accomplish anything significant at all, we would have to plan on no less than seven working days underground.

"Joe, you mean these thirty people are going to drag all their food and equipment into the cave before they start working?" said Bloch. "Why, it'll take two or three days just to complete the build-up in Camp One. Then you'll have a group of exhausted malcontents on your hands. Never heard of a cave mutiny before, but we may have one yet!"

Bloch had thrown a stumper at me. Parker and I thought in silence.

"What about this? Why don't we drill a well, maybe six inches in diameter, down to the Lost Passage?" said Parker. "Then all we have to do is dump in six cases of assorted canned food every morning."

"Fine idea," I said. "But we don't even know where the Lost Passage is. It might be two miles away, then again it might be right under Grand Canyon. That's what we're trying to find out."

More silence.

Bloch had an idea. "We'll have to think in terms of more people. That's all there is to it. Since we don't know the magnitude of what we're getting into, let's face it; all we can do is set an arbitrary minimum figure of thirty people for a week underground. From that constant, we have to figure out how many people we need to supply them." The foot was in the door. There would be more people.

In the United States up to this time caves had been studied by small groups of people on short trips. Week-ends found cave explorers journeying by automobile to a cave, perhaps a hundred miles away, then returning to their jobs on Monday mornings. It's an ideal way to study caves since the problems of large-scale expeditions are avoided. Consequently we were beginning to see the inadequacy of our experience in dealing with problems growing out of a cave too large to be handled by conventional means. We must have a large expedition to pry loose the secrets buried beyond the Crawlway. Finding no ready answer to the exact size of the expedition roster, we decided that we might determine the optimum proportions by starting personnel planning from the top. We agreed to follow the military concept of separating command and staff functions.

Under me, at the top of the command pyramid, would be two assistant expedition leaders. On the next level there would be a support group leader at the Base Camp on the surface directing supply operations. There would be an exploration group leader at our main underground camp directing operations there. All work in the cave would be performed by small parties of from two to four people, since experience in smaller caves had taught us that larger parties are slow and inefficient. Safety considerations make exploring by lone individuals out of the question. Each party would be commanded by a leader appointed in turn by the support or exploration group leader.

The specialists who made up the staff were to be a geologist, meteorologist and hydrologist, biologist, chief surveyor, chief photographer, and physician. The staff would also include those responsible for the technical aspects of support: a communications officer, supply officer, and information officer.

The table of organization appeared complete on paper, but the names of individuals who would fill those slots would determine the success or failure of the expedition. Slowly names were lettered in the proper boxes. The first selected was John L. Spence, a design engineer and president of the New York Camera Club. His background in the development of specialized photographic equipment, coupled with his extensive photographic knowledge, made him a logical choice for chief photographer. He, as all those who would be chosen, had extensive experience in caves of varying degrees of difficulty. A pictorial record of the expedition would be essential, both for the scientific aims, and for the documentation of its progress. After his appointment, he made frequent trips from New York to Philadelphia to participate in planning sessions.

For a second-in-command I wanted someone thoroughly familiar with the known portions of Crystal Cave. Bill Austin was the ideal man for the job, so I lateraled the ball to him and we both started running. On the choice of third-in-command I picked William Stephenson, pioneer in American speleology and founding president of The National Speleological Society. He brought with him a wealth of leadership experience garnered from almost fifteen years of leading cave explorations.

In response to a story about our activities in *The News*, the monthly publication of the N.S.S., applications poured in from experienced members of the society scattered around the country. Even non-members wrote for application blanks. One girl rodeo rider of adventuring blood pleaded to go along, even if she would only be allowed to cook inside the cave. She claimed valuable experience from leading Girl Scout groups on camping trips, but admitted that she had never attempted cave exploring. She had done almost everything else: driving race cars, climbing mountains, and skin diving. She had done everything, it seems, except navigate Niagara Falls in a barrel. I had to tell her that she could not be considered for the expedition. We had no place for daredevils.

In the vanguard of applications submitted were the names of several men already experienced in Crystal. Roy Charlton, the lean, wiry farmer from Virginia who had been with me on the three-day trip to the lower levels, was eager to return. Bob Handley, a lanky West Virginian, had taken "the tour" in Crystal's lower levels several times. Applications from Ohio came in simultaneously from Roger Brucker, Phil Smith, Roger McClure, and of course Jim Dyer. Austin contacted Jack Lehrberger in Louisville, a man well-versed in the problems of cave exploring in the "big cave country" of Kentucky. An application was there from Luther Miller, a veteran of earlier assaults on the endurance barrier in Crystal's lower levels.

Applications from the cream of the crop of American cave explorers filtered into the society's headquarters in Washington, D.C. Ackie Loyd, an experienced hand in Virginia caves and an extremely intelligent observer of the underground world, wanted to go along. Ida Sawtelle's application was there. A dog trainer by profession, she was acknowledged one of the nation's outstanding women cave explorers, having ventured into Schoolhouse Cave, the country's classic underground rock-climbing challenge. Earl Thierry, civil engineer and part-time cave explorer, had caved in West Virginia, Pennsylvania, and Virginia. George Welsh knew the tougher caves in Indiana and Kentucky.

There was an application from Dr. Ralph W. Stone, former State Geologist of Pennsylvania, and one of the nation's outstanding speleologists. He wanted to help in any way he could. I was overjoyed that he wanted to serve as advisor for the geologists.

Screening the applications and considering each man on the basis of his experience, and personal observation by myself and others, proved to be a difficult and continuing job which would not be complete until the final hour when the expedition assembled at the cave.

We had arrived at a date for the expedition; we would start into the cave on Sunday, February 14th, 1954, and emerge on the following Saturday, the 20th.

Working continuously on the personnel selection problem, the planning staffs also considered the commonplace. How would explorers stand sleeping underground night after night? We had already proved that a few of us could take it for three days, and the French in Padirac had been able to extend their stay to six. We also knew that at the end of that time both groups had nearly reached the breaking point in terms of exhaustion. When one of the explorers who had been with us on our three-day trip into Crystal heard of our discussions, he wrote: "I don't much like the idea of camping underground for more than two nights straight. Somehow it dampens the spirit. . . ." Soggy spirit or not, we decided to take the calculated risk and try it.

Transporting equipment through the Crawlway, and especially through the Keyhole, loomed in our minds as a critical problem. We knew a sleeping bag would go through the Keyhole's 10-by-24-inch opening, but a constant stream of supplies of every description was another matter. Several of us lost a few good night's sleep over that puzzle, but remembering that our chief photographer, John Spence, was a design engineer, we let him worry about that problem.

We considered the problems of feeding thirty people underground on a sustained basis. For cooking we had momentarily considered dragging bags of charcoal down to our camp, but we thought better of it. Profiting by previous experience in the cave, we remembered that meals had been prepared for many years at Floyd's Kitchen in the Lost Passage on a Coleman gasoline stove. Pound for pound, we would get more heat from gasoline, but we didn't reckon fully with some problems that were to be an outgrowth of this decision.

Working together, we drew up a list of personal clothing and equipment to be carried in by explorers and scientists based at the underground camp. We relied on my experience in Crystal and the experience of others in smaller caves for the completeness of this list. We planned to purchase large vinyl plastic bags to issue to each man in which he could pack his gear, reasoning that the bag would keep out moisture and insure a dry change of clothes.

Continually we lapsed back, finding ourselves thinking of Crystal Cave as just another cave of the variety we habitually visited. Communications were no problem in these, but operations in Crystal would be another matter. The size of the group, the distances and the logistics needed, demanded that we be able to communicate between underground camps and Base Camp. Charles Mohr and I had talked of a telephone line, but neither of us realized that, far beyond a convenience, it would be a necessity. Without telephone communications we might just as well fold up our tents and go home. Survey notes and explorers' findings would have to be conveyed to base camp instantaneously, so others could plot and evaluate them. No phone meant no correlation of data until it would be too late in the week to be of use to continuing operations. We had to know at all times the status of provisions in the underground camps, and inventory depletions would have to be replaced as soon after they occurred as possible. We would have many simultaneous activities going on in various parts of the cave and, without coordination, it would become a three-ring circus with no meaning. Jack Parker brought up the subject of the value of a telephone system in the event of an accident, when the efforts of rescue teams would have to be planned. His arguments clinched the case for a phone system. Bob Lutz, our communications officer, had a job on his hands.

To maintain a permanent record of the expedition's activities we purchased a series of logbooks in which data would be recorded. Each camp would have a log, and in it would be entered the arrival and departure times of each party, the gist of important phone calls, and any findings that ought to be preserved. The information officer above ground would use a tape recorder wired into the telephone system to record important conversations and oral reports.

Whenever an attempt is made to do something that no one has tried before, the press and radio become interested. At our press conference in Washington, D.C., on January 9th, 1954, in which we announced our intent to the public, we became suddenly aware of the impact of our plans. Ivan T. Sanderson, vice president in charge of N.S.S. public relations, made a quick trip to Philadelphia to discuss with me what we ought to do to satisfy public curiosity. Obviously we had to make some provision to accommodate the working press. He pointed out to me that we could not possibly let every reporter who asked go into the cave. I heartily agreed, for the treacherous wilderness beyond Scotchman's Trap was no place for a novice. Ivan advanced the following plan: One newspaper reporter representing a pool of wire services, one photographer representing a national magazine, and one radio reporter representing a nation-wide network would be allowed to accompany us underground. This plan seemed to satisfy our obligation to the public with a minimum of burden on the expedition, so Ivan scurried to Washington to iron out the details with the expedition's information officer, Burton Faust.

As more staff officers were appointed, I too found it necessary to scurry to Washington for planning meetings. At these, we delved in detail into the manifest problems of supply, communications, and medical considerations.

When expedition physician Halvard Wanger, M.D., was chosen to fill the medical post, he charged into action, sending lengthy mimeographed sheets to everyone going. He prescribed immunization shots and drew up an elaborate list of calisthenics aimed at getting people into excellent physical condition for the ordeal they would be going through later. Within two days after the calisthenics list had reached members painful groans rent the air: the agonized shrieks of cave explorers with sore muscles.

Planning shifted from Philadelphia to the staff officers as soon as they were appointed, then I began receiving requests from those officers for assistants. The information officer needed stenographic help at the cave. The communications officer needed linemen and switchboard operators. Tom Barr, Jr., who was by this time in charge of both biology and geology studies, recommended two people for the posts he had created: assistant biologist and assistant geologist.

Dr. Wanger was active in this conspiracy. One of his frequent telephone calls from Shepherdstown, West Virginia, went like this:

WANGER: Are you going to have a nurse on the expedition?
LAWRENCE: A nurse?
WANGER: Well, I'll need some help. We ought to have a nurse along.
LAWRENCE: Look, Doc, if anyone gets that sick we'll send them to a hospital. We're not going to run one at the cave.
WANGER: You want me to study the physical effects this trip has on people? This means thorough physical and psychological examinations.
LAWRENCE: O.K., Doc. I'll keep my eyes open, but I'm not too sure I can find a nurse who wants to go.

Switchboard operators, stenographers, nurses! The magnitude of the operation was increasing. What next, assistants to the assistants? I soon found out. My wife asked me how many cooks I planned to have. When I told her plans called for one at each camp, she raised her forearm to her brow and sighed. On learning from her that each cook would need one or two assistants, I hit the ceiling. I could see visions of two or three hundred people milling around supporting each other; a laundry, a filling station, a cafeteria, a repair shop. . . . Taking great pains to appear cool and calm, I said, "There will be *one* cook at each camp, and that's final!"

By letter Doc Wanger informed me that he had found two excellent nurse prospects, either one of which would fill the bill. In the next mail I received a letter saying they had *both* accepted and weren't we lucky to have them! Then another phone call. This time *I* decided to worry *him*.

LAWRENCE: Say, Doc, could you use an assistant physician?

WANGER: What? Well, I don't know. I have these two nurses . . .

LAWRENCE: There are three physicians, interns at Philadelphia General who want to come. They're good cavers in excellent physical shape.

WANGER: Three of them?

LAWRENCE: Yes, three. Can you use them?

WANGER: Not three! Now wait a minute. I might have a little something for them to do . . . Be good experience for them. They'd be good to have in case of an accident. But I can't keep them all busy.

LAWRENCE: O.K., Doc, I'll put them on the roster anyway. My wife says I'll need some K.P.'s.

Gradually order developed out of chaos. With work delegated to staff members we had stamped the seal of approval on the blueprint for action. All that remained was to follow the plans.

4

STOCKPILE

The selection of equipment confronted the planners with a myriad of details. We must prescribe personal equipment for those in the cave. We must select equipment for exploring parties, for underground camps, and for a surface camp. We would need emergency rescue equipment, communication equipment, medical supplies, and scientific instruments. All equipment that went into the cave had to go through the Crawlway. This placed rigid requirements of lightness, compactness, and sturdiness on our selections, and we would take in only what was absolutely necessary.

To solve the logistical problem posed by the Crawlway, John Spence undertook the development of a metal canister for transporting supplies. He and Russ Gurnee designed and built four canisters for use in the Crawlway. Working in Gurnee's sheet-metal shop, they placed a metal cone on the forward end of the tapered, galvanized steel cylinder. It could be dragged by a rope tied to a ring at the point of the cone and would not snag. A canvas cover closed the rear end after it was loaded.

Although we wanted to sleep comfortably at our underground camp, we wanted to transport only a minimum of sleeping equipment through the Crawlway. On our three-day trip in 1951, we had taken in half as many sleeping bags as people, taking turns using them. We decided to follow this plan on our week-long trip. Since it was very important that we get adequate rest if we were to stay underground for a full week, we added air mattresses to our equipment lists so that we would be as comfortable as possible. For sanitary reasons, we planned to issue each individual his own lightweight, cotton sleeping-bag liner.

One member attempted to solve the problem of sleeping in a damp cave by constructing a lightweight hammock. With the ends secured to formations, he argued, it would suspend him safely and comfortably over the damp floor. Every night for a full week he knotted nylon ropes, and his labors were approved, if somewhat skeptically, by all who saw the results.

The cave's dampness posed another problem in supply. The explorers would invariably get wet while exploring among the waterfalls, streams and pools that are so common underground. Humidity in caves often approaches 100 percent, so wet clothing is difficult to dry. Cold, wet clothing can be one of the greatest detriments to high morale underground. The use of desiccants seemed the solution to this problem. By placing desiccant beads in our plastic clothing bags, we could dry soggy cave clothes.

We obtained samples of several types of desiccants and made some calculations. Wet clothing that has been wrung out will weigh from one and a half to over two times its dry weight. The desiccant we considered would absorb no more than its own weight in water. Therefore, the weight of desiccant required to dry our wet clothing would be quite close to the weight of replacement items. On the basis of this analysis, there seemed to be no real justification for the use of desiccants. Instead, we would bring in dry clothes as needed.

We included a one-man rubber life raft so that we would be able to explore any large underground rivers we might find. For difficult rigging jobs or for rescue work we assembled a kit of heavy tools including pulleys, rope, a star drill, jacks, and a crowbar.

Each staff officer was responsible for assembling the specialized equipment he would need. For communications there were telephones, wires, and switchboards. The doctor had medicines, bandages, and plasma. The biologist needed hundreds of specimen bottles while the information officer procured a tape recorder and three typewriters.

The photographers needed equipment not available on the commercial market. John Spence, the expedition's chief photographer, built a strong, compact box to hold a flash unit and two cameras—one for black-and-white and one for color pictures. The cameras could be operated without removing them from the box by shooting through a Plexiglas window. Spence also constructed a box to protect the supply of fragile flashbulbs.

Because Bob Halmi, photographer for *True* magazine, planned to take thousands of pictures underground, flashbulbs were out of the question. He sought the aid of a manufacturer of stroboscopic equipment for help on his lighting problem. Up to this time there was no electronic flash unit strong enough and lightweight enough for Halmi's needs. The firm's engineers went to work on the problem and soon developed a strob unit that met Halmi's requirements for ruggedness, compactness, moisture resistance, and prolonged life. Three were built for the expedition.

Compasses, tapes, and altimeters were needed for surveying. Earl Thierry, our chief surveyor, had several compasses and tapes, but we needed more. Phil Smith and Roger McClure went to the Ohio Division of Geological Survey in Columbus for help. Could we borrow a compass, a Brunton compass? The answer was yes. Smith and McClure left with two compasses and three altimeters. Roger Brucker procured a fourth altimeter for the use of the expedition.

Facilities in our Base Camp on the surface must be identified. Brucker painted signs for the mess hall, infirmary, supply, and communications tents. Remembering that a rock-climbing team from France had been invited, he included a picture of a spoon on the mess hall sign, a telephone on the communications sign, and so on. Unfortunately, the Frenchmen were unable to join us even though they had planned up to the last minute to participate. The pictures, however, made the signs more attractive.

We drew up a list of recommended equipment that we sent to all expedition members. Combining our recommendations with their own experience, they selected their gear. Most chose coveralls for their outer garment. The one-piece garment was judged superior to two-piece clothing because there was less possibility for it to snag or bunch up on the way through the Crawlway.

Hard hats, such as miners wear, were mandatory. They offered protection from rocks that might be dislodged above an explorer. They also prevent injury if an explorer bumps his head on a low, irregular ceiling. There are cases on record where helmetless cavers have received scalp cuts requiring several stitches.

Some persons, remembering the pits in Crystal, thought it absolutely necessary that the hard hats be clearly visible if viewed from above, so no one would toss a rock into a pit to determine its depth if a hat at the bottom shone with reflective tape. Scotchlite reflective tape strips that could be seen from a distance in a dark cave were put on many of the hats. Some put strips on all of their equipment so that it would be easier to find.

A caver's most important piece of equipment is his carbide lamp worn on the hat, leaving both hands free for climbing and crawling. Since the lamp is clamped on the front of the hat, it turns with his head, always casting its beam where the caver is looking.

Each explorer carries a supply of carbide and water for recharging his lamp. The containers for these items have to be small and compact, for each item of protruding equipment makes crawling more difficult. Individuals have strong opinions about containers. Some choose metal boxes specifically manufactured for carbide containers. Others select tobacco cans, tobacco pouches, or small plastic boxes the Navy has used for first-aid kits. Army canteens are selected by some for carrying water. Others, who wish to avoid the clumsy bulk of these canteens, choose plastic bottles. Roger Brucker uses a plastic hip flask of the variety used by celebrating alumni at football games. It is curved to fit the hip and is flexible.

A cave explorer is helpless without light. If his light fails he cannot move without risking injury. Each explorer was instructed to bring a flashlight as an emergency source of light. The flashlight would have to be sturdy to stand up under such strenuous exploring. It was also desirable that it be waterproof. Not only would the flashlight serve as an emergency light source; it would also supplement the carbide light. The carbide lamp does not cast a spot pattern but provides a flood of light illuminating a wide angle. For this reason, it lacks the penetrating power and brightness of a spotlight. When the explorer wanted to inspect a pit, or a long passage,

An explorer's equipment includes hard hat and carbide lamp, gloves, knee-pads, containers for carbide and water, flashlight, perhaps candle and matches, and rations. *Photo by John Spence.*

his small-angle flashlight beam could reach well beyond the limits of his head-lamp.

Knee protection was a necessity. The long crawlways of Crystal Cave would abrade unprotected knees into painful uselessness. A durable rubber knee-pad of the type used by miners and gardeners was selected. This particular item was foreign to the inventory of most cave explorers' equipment, since few other caves in the country require crawling for such great distances.

Shoes would have to be heavy, high-top work shoes, army shoes, or climbing boots. Plain leather soles could not be used, for they slip on mud and wet rock. Good traction is demanded for straddling canyons and climbing pits. There was a wide variety in the selection of shoe soles. Some chose plain rubber, others, cleated rubber, and still others preferred soles with spikes and nails. There were tricuni nails, golf spikes, and hobnails. Each explorer would expound long and loudly the advantages of his own footgear. Some types were better than others for a specific job, but where one excelled in one way, another was proven better in a different situation. In

Climbing equipment includes rope, pitons, karabiners, piton hammer, boots with nails or cleats, and a flexible metal ladder. *Photo by John Spence.*

the final analysis we found that all forms selected were adequate for the trip.

Rock climber Lou Lutz already owned tricuni boots and a wide selection of additional climbing gear. He had pitons in a variety of sizes and shapes, and a piton hammer. A piton is a spike that climbers drive into cracks in the rock to obtain an anchor point for a belay. He had karabiners—snap rings that are used to fasten a rope to a piton. Lou insisted on new climbing rope so that he could be sure of its strength. First he and the other rock climbers pulled it tight, working the kinks out of it. Then they climbed with it, working out any remaining kinks and "breaking it in." Finally, Lou tied his rope in neat coils, loaded his climbing gear into John Spence's car alongside cameras and flashbulbs, and the two started the journey to Crystal Cave. The migration had begun.

5

UNDER CANVAS

FROM ALL OVER THE COUNTRY cavers converged on Floyd Collins Crystal Cave. Roger Brucker and Jim Dyer were the first to arrive. Tuesday morning, five days before the expedition was scheduled to go underground, a gray dawn broke as their station wagon stopped in front of Austin's house. They knew Bill Austin would still be in bed, so they crept into the winter office, unrolled their sleeping bags, and went to sleep.

A short time later Bill burst through the door. "Time for breakfast, Brucker," he yelled. Through one eye and then another, Roger saw that Jim was already up.

Over a breakfast of sausage and eggs prepared by Bill's wife, they listened to the latest plans for the expedition. The communications team was supposed to arrive Friday afternoon and begin stringing wire immediately.

"Say, Roger, did you hear that plan for running field telephone wires all the way down to the Lost Passage?" said Bill.

"A real panic," he answered. "It looks to me as though the communications officer has never been in Crystal."

"We'll be lucky to get *one* line down. It's hard enough to get there, let alone drag wire after you," said Bill. "I hear he's bringing *eight miles* of wire with him!"

A car crunched on the gravel outside. They went out to find a man wearing a reporter-type hat. He introduced himself as Ed Easterly, District Chief of the Associated Press.

"I've come down here," he said, "to get some background material on the expedition." With that he took out his notebook and began to write. Without talking very long they realized that Mr. Easterly needed a glossary of cave terms. Brucker wrote out some definitions for him.

Then Easterly got down to business. "Just how far can you go in the cave?" he asked.

"We don't know," said Bill. "We've never reached the end."

"Well, let me put it this way: how far have you gone in previous expeditions?"

"Again we don't know," said Bill. "We've never measured it." Outside of the commercial portion of the cave, no one in the world had any idea how big it was; nobody who had ever been in Crystal knew how far he had traveled. This much was known; it took about twelve hours to get from the entrance to Bogardus Waterfall. From that point there were literally dozens of passages that previous explorers had never entered. It might take twelve hours or a hundred and twelve to run some of the passages out to their ends.

"But what about the statement on the folder?" said Easterly. "It says there's an estimated forty miles of cave altogether."

"*Estimated* is the key to that," said Jim. "Actually no one really knows what's down there. That's why the expedition. Sure, we know a great deal about how to get in and out, and we've done some exploring, but for every place we've gone, there must be ten we never looked into."

"Perhaps I can help," said Brucker. "I've only made one trip in, but I can see what you're driving at. You seem to think this cave is like one long tunnel, and all you have to do is walk to the end of it. Actually, it's like a series of four or five spider webs superimposed on each other. The webs interconnect at many places, and that makes the whole thing more confusing."

"And," added Bill, "we don't know yet how far the extremities of these spider webs reach."

Ed was beginning to see the magnitude of the job confronting the expedition. He asked a few more questions, then asked to see the commercial part of the cave. Bill Austin nodded approval, so Brucker acted as guide.

Ed was greatly impressed at the size of Grand Canyon as the lights came on. They continued, past the Valley of Decision through Devil's Kitchen and along the Gypsum Route.

"How far have we come?" he asked.

"About a mile," said Brucker. "We're almost to Scotchman's Trap where the expedition's route to the Lost Passage starts."

"I see some of the supply problems already," Ed said. "It's a long walk back here."

They were at the end of the string of electric lights now. They switched on flashlights and walked for several minutes to a point where a huge boulder appeared to block the path. They ducked under it and up a mound of yellow clay on the other side. "End of the line," Brucker said. "All out for Scotchman's Trap."

Easterly shone his flashlight beneath the boulder leaning against the wall. He probed the light downward into the vertex of a cone. The light picked out a hole about eighteen inches in diameter. Everything the explorers would use, every sleeping bag, every morsel of food, every mile of wire would have to go through that small opening.

He looked at Brucker in amazement, then he looked back. "Why do they call it Scotchman's Trap?" he asked.

"Because it's so tight," came the pat answer, one of Jim Dyer's favorite "tourist" jokes.

At lunch Ed Easterly asked why people went cave exploring. Austin, Dyer, and Brucker earnestly tried to explain, but the discussion eventually ended in the telling of one funny cave-exploring story after another, such as the time Bill Austin jumped into the rubber boat in Hidden River Cave with his spiked boots on.

After lunch Easterly wanted to know what was available in the way of phone lines. None ran to the cave, and the closest phone was located in a national park fire tower about half a mile away. It was a private phone, part of the park's communication system. Easterly wondered if arrangements had been made to hook into it, and Bill said they hadn't. Easterly went to see Park Superintendent T. C. Miller who granted permission to tie into the line providing it would be kept open in the case of a fire report.

Robert J. Richter arrived in the late afternoon. As he unloaded his gear from the trunk of his car, Dyer and Austin noticed two large shiny aluminum boxes.

"What's in the boxes?" asked Jim.

"Soil sample bottles," he replied, "part of the scientific program. We intend to collect soil samples from all parts of the cave, then send them to Chas. Pfizer & Company—a large pharmaceutical firm in New York. They'll analyze the soil samples for antibiotics. Who knows, maybe we'll discover another mold, like penicillin."

"You mean you're going to try to take those boxes through the Crawlway?" asked Bill Austin in amazement.

"I won't. The supply team will. That's their job."

They looked at each other, then at Richter. "Those boxes won't fit through the Crawlway," said Brucker. "They measure about fourteen inches square and about twenty-four inches long. The Keyhole would stop them if nothing else would."

"I thought this was supposed to be a big cave!" Richter said. "What do you mean, you can't get the boxes through?"

"The Keyhole is the limiting factor in the whole supply set-up," explained Bill. "It's the tightest part of the Crawlway—a place where you have to transfer from one passage into a parallel passage. The hole measures less than ten inches high and about two feet wide."

"Then I guess we'll have to take the bottles in loose," he said.

Austin and Brucker left Richter counting bottles. They went into the town of Horse Cave to pick up a sink and stove for the mess hall. The carpenters had been working all week nailing a floor and walls into Floyd Collins' old summer office that was to be our mess hall. When they returned, the last carpenter had finished. Brucker borrowed two of the cave's tourist guides and set to work making a counter for the sink. By five in the afternoon they completed this and started to work on a table from which expedition members would eat standing up. As they worked, loads of food and supplies were carried in and deposited on the floor, generally in their way. The kitchen was ready for business by 9:00 P.M., so Brucker went to find Jim Dyer.

He walked out of the door of the summer office and saw lights glaring in the medical tent. Doctor Wanger was busy setting up medical supplies, a hundred containers full of assorted pills and liquids.

Like Dyer and Brucker, I too traveled all night to reach the cave area. Early Thursday morning Hugh Stout, Albertine Talis, Tom Barr, and I parked in front of a hotel in Cave City, registered, then disappeared for a few hours of sleep.

When I came down into the lobby after a brief nap a New York caver greeted me. "Hey, Joe, glad to see you. There are some important people here I want you to meet."

He ushered me through the lobby, introducing me to some local business men he had met the evening before. Next he said, "Let's go across the street to the Bank. I want to introduce you to the vice president. You should also meet the Judge. I had a long talk with both of them last night."

When we returned to the hotel from our visits, the desk clerk said, "Doctor, there was a gentleman in here looking for you. He wanted to show you some blind fish." This startled me. It was the first time I had heard this New Yorker referred to as "Doctor." Actually, he was still working on his bachelor's degree.

After learning what a politician my friend was, I decided to put his talent to work. " 'Doctor,' you are authorized the cost of one long-distance telephone call to contact the home office of Pillsbury-Ballard and ask them if they would like to contribute a supply of canned biscuits to the expedition."

A couple of days later I walked into the expedition's Base Camp mess hall and saw some cardboard cartons stacked against the wall. "What's that?" I asked.

"One-hundred and forty-four cans of Ballard's Biscuits," answered the "Doctor." "They just arrived by express, prepaid."

As more and more expedition members arrive, Austin's parking lot fills up. Buildings are, left to right, Austin's home and office, the Collins' old homestead, and the summer office converted into a mess hall for the expedition. The cave entrance is out of sight 100 yards behind the summer office. *Photo by John Spence.*

Actually, a number of American manufacturers answered our requests for help, usually made further in advance. They contributed generously of their products: apples, flashlight batteries, carbide, rope, and many other items were made available to us.

Ida Sawtelle arrived from Brooklyn on Friday, two days before the expedition was to go underground. I spotted her in the hotel lobby in Cave City.

"Ida, you're on K.P."

"But I haven't got to the cave yet," she protested.

"You'll be there in another fifteen miles," I said. "Your job is to put the mess hall in operation and cook through Sunday. The plumbing is being installed now."

"There were some things I wanted to do first," said Ida, "but if you need me now, I'm ready to go to work."

"You have plenty of time. You don't have to serve until supper tonight, but plan to feed thirty people then. I'll give you three helpers. Alden Snell, the supply officer, will be your boss."

Three helpers were soon selected for Ida. When Audrey Blakesley, a rock climber from Trenton, New Jersey, breathed too hearty a sigh of relief because she had missed K.P., I said, "Get yourself a shorthand pad, Audrey, you will be my stenographer for the next two days."

Friday, tents went up as Base Camp was established at the cave entrance. Aside from the single frame building that we were using for a mess hall, everything else had to be housed under canvas. Dr. Wanger's infirmary was in a large tent. Another tent went up for my office. The largest tent was erected to house communications and Base Camp headquarters. Still another tent was used for supply. Others went up for sleeping quarters. The tents were wired for light and electric heaters. By Thursday night, Base Camp was a reality under the stars of a February sky.

The expedition's tent city goes up in a windy field close to the cave's entrance. *Photo by Robert Halmi.*

Part II

PENETRATING THE BARRIER

by Roger W. Brucker

6

THE DOCTOR TAKES A TOUR

BEING A MEMBER OF AN EXPEDITION was a totally new experience for me. Back in Ohio a group of us had journeyed to nearby caves in Indiana, and we had explored the crawlways of our own state. When the Air Force had stationed my unit in New York City, I had the good fortune of exploring caves in Pennsylvania, New Jersey, Connecticut, and upstate New York with some of the people who were now milling around Base Camp. Phil Smith and Roger McClure walked by carrying boxes. I remembered the first trip we had taken together to the lower levels of Crystal four months ago, and how that visit had formed our determination to return.

As I strolled among the tents filled with workers, each doing his assigned task, I met Dr. Wanger with a hammer in his hand. He told me that he wanted to drive some more stakes in the ground so his infirmary wouldn't blow away. When I asked him where Joe Lawrence was he pointed toward one of the white buildings.

I walked into the winter office to find Dyer, Austin, and Lawrence discussing tomorrow's plans. "Brucker," said Joe, "how would you like to take the Doctor to the lower levels tomorrow?"

"I don't know the way. I've only been in once before," I said.

"Jim Dyer will lead. We want you along in case anything unfortunate should happen."

In case anything happens, I thought to myself, what good would I be? I didn't know the way in, let alone out. Besides, what could happen? The more I thought about it the more I thought that perhaps something could happen. I didn't know Dr. Wanger then, and I had no idea how much

51

punishment he could stand. Suppose he fell into a pit. Suppose he became fatigued and couldn't go any farther. Suppose. . . .

"I'll go," I said. Joe explained that there were three purposes for the trip. He wanted us to locate a campsite near a good supply of drinking water; he wanted to give the Doctor a taste of what expedition members would be going through; and he wanted the Doctor to determine for himself the feasibility of removing an injured man from the depths of the cave. We added a fourth objective when we decided to carry in two sleeping bags. This would give all of us a personal insight into how supply teams would stand up under the rigors of travel in Crystal Cave.

Would Crystal be as confusing to me now as it had been the last time? I didn't have to wait long to find the answer.

Friday, February 12th, dawned sunny and bright. Outside my tent people were hurrying back and forth, carrying supplies. At the mess hall there was a breakfast of bacon and eggs served among swarms of people who stood at two long tables. Nancy Rogers and Ida Sawtelle were chief cooks, dishing out fruit juice and toast. Excitement seemed electric. Some of the members who had spent the night sleeping in Grand Canyon in the cave, instead of in tents, described their experience; sleeping was good, they reported, but it was disconcerting to wake up and find yourself in total darkness. One member mislaid his flashlight and had to wake up a friend to help find it.

I went outside to find Joe talking to Burton Faust and learned that we would go in as soon as everyone was ready. Not wanting to hold things up, I went to the car and climbed into my coveralls, loaded my plastic box with extra carbide, and filled the hip flask with water. The Eveready flashlight worked, but a couple of candles went in my pocket for good measure. With my knee crawlers fitted neatly on my rope waist band, I was ready to go.

Jim Dyer was talking to Bill Austin when I approached. He said he'd be ready in a jiffy. Dr. Wanger stood by his car. "What do I wear in this cave?" he asked. I pointed out the various items of my equipment, which he mentally listed, then he turned to dig a couple of other things out of his car trunk.

I wandered over to the mess hall to see about getting a lunch to eat down below. Ida said she'd make one.

Roger McClure and several others were rolling sleeping bags to go to Camp One down in the Lost Passage. "Tighter," yelled McClure, "These have to be as small as we can make them." McClure spoke from experience. He demonstrated how he wanted them rolled; he unzipped a bag and placed a rubber air mattress in it, an inflatable pillow, and a sleeping bag liner. The bags themselves were stuffed with kapok and lined with red cotton flannel. He zipped the bag closed and folded in the sides slightly. I pitched in at that point and kneeled on the top end of one bag while he rolled from the bottom, tightening the roll with all his strength. The bag's protective cover furled around the roll and the draw strings were tied securely. "There," said McClure, "That ought to go through in one piece." It would have been impossible to roll the sleeping bags into tighter bundles by hand, and the result was a roll about nine inches in diameter, and 24 inches long. I picked up two of the rolled bags and headed for the Doctor.

In less than twelve hours a row of larger tents flanks the nerve-center at Lawrence's tent headquarters. *Photo by Henry Douglas.*

Doc Wanger was getting his sleeves tied with string because he didn't intend to take half of the sand in the Crawlway along with him. His son, Bill Wanger, finished the last knot.

"Do I need this?" Doc asked. He held up a bundle of nylon rope.

"No, don't take it," I said, remembering how a rope I had taken on my first trip had snagged on the rough walls of the Crawlway. "Jim will be leading us by a route where we won't need rope," I said. Doc tossed the rope back into the trunk of his car.

Jim joined us, saying Ida had the lunch ready. I went to pick it up. Cheese sandwiches, dried fruit, peanuts; not very hearty fare for Crystal crawlers. I stuffed it into my rucksack. We were ready.

The three of us said goodbye to Bill Stephenson and Joe Lawrence. Doc slung a sleeping bag over his shoulder and fell in behind Jim, who led the way down the hill to the entrance. I looked up at the sky as we descended, knowing I wouldn't see it for at least twelve hours. It was bright blue against the wires which furnished current for the lights in the commercial portion of the cave. We closed the door behind us and we were in blackness. Jim flicked a switch and the lights went on. About fifty feet down the passage we stopped to fill our pockets with apples. Four bushels of Shenandoah Valley apples were piled there for incoming and outgoing expedition members; we bit into them as we marched down the canyon trail into the cave.

At Scotchman's Trap we sat down and rested while everyone filled and lighted carbide lamps. Doc Wanger pulled up his pants legs.

"What are you doing?" I asked. He waved some strange looking elastic bands in my face.

"Basketball knee-pads . . ." he said, slipping them on snugly around his knees.

"They look a little thin," I said.

"They'll work all right," he said. "Besides, they won't be as clumsy as those big rubber ones. I don't usually wear knee-pads in caves anyway."

I had heard those words before. He pulled down his pants legs, then raised his sleeves and proceeded to put on elbow-pads. We had an audience by this time. Expedition members had come to watch the send-off, augmented by press photographers, who kept asking for "just one more."

"Let's go," said Jim quietly. He scrambled down the slide under the boulder and disappeared through the small hole that was Scotchman's Trap. Doc tossed his sleeping bag down the hole and piled in after it. He jammed it through the hole and took a last look at the people ringing the upper rim. Then he was gone.

It was noon by my watch as I threw my sleeping bag down the hole and followed. The apple tasted delicious; no point in throwing it away until we arrived at the tough part of the crawl. Jim and Doc were ahead, so I decided to wait until they were almost to the cross-canyon before I followed. That way I wouldn't have to stop frequently in cramped quarters. Doc threw the sleeping bag ahead of him, then crawled up to it and repeated the process; I could always tell if he was moving ahead by the "thump, scrape, scrape, thump, scrape, scrape" of the bag hitting the floor and the Doc crawling. Jim was out of sight now, so I began to crawl, hurling the bag as far as I could in the cramped quarters. It traveled about ten feet. In a few moments I had squeezed down into the narrow canyon and was following on the heels of Jim and the Doctor. Gypsum lining the walls of the canyon glistened under the carbide lamps like a million tiny diamonds. We held the bags over our heads now and moved along sideways. It looked familiar. Doc was sweating profusely as he strained and pushed his way through. I remembered how I had perspired on my first trip in and was amazed to discover how cool I felt now. The only way I could account for it was that, having been through before, I knew how to travel expending the least energy.

The passage was lower now as we came into the small room called Last Chance and stood up. Doc panted and rubbed his face with a handkerchief. Jim looked cool and ready to go, the way I felt. "Tell me," said Doc, "is all of the Crawlway this rough going?"

Jim looked at me. Mentally we went down the list of obstacles yet to be encountered, the S Curve, the Keyhole, Straddle Canyon. "A good bit of it is rougher than this," said Jim in the understatement of the trip. We moved on, this time on our hands and knees. I knew Doc was wondering what he might run into up ahead; so was I. Would the cave prove to be as confusing as the last time I went through it? So far, it was familiar.

A muffled voice came from the passage ahead. It was Jim, yelling something about the S Curve. We were literally wiggling along on our stomachs, inching the sleeping bags ahead. Now and then Doc tried to adjust his basketball knee-pads, which kept slipping down his leg and offering no protection.

"Roger, how do I get through this?" he asked. I couldn't see around him since his whole body plugged the passage.

"Push the sleeping bag hard, then hold on!" I yelled. I knew for certain he was at the S Curve now, so I rolled over on my back and rested with my nose a scant three inches from the ceiling, listening to the struggle ahead— a man trying to push himself where men weren't made to go. He grunted, swore, then rested a moment and tried again. The ceiling was terribly close, I thought, what would happen if it caved in? Either here or back farther toward the Trap? They would come looking for us after fifteen hours or so, and what would they do when they crawled upon that awful pile of stone? It couldn't be moved, except out of the Trap piece by piece. They really couldn't drill down to rescue us if we were trapped, for beyond the Trap no one knew how the passages lay. There was a map on the old commercial tourist folder which showed the Crawlway, but that was inaccurate. It couldn't be of any practical use to a rescue party. I thought about it in silence. Silence!

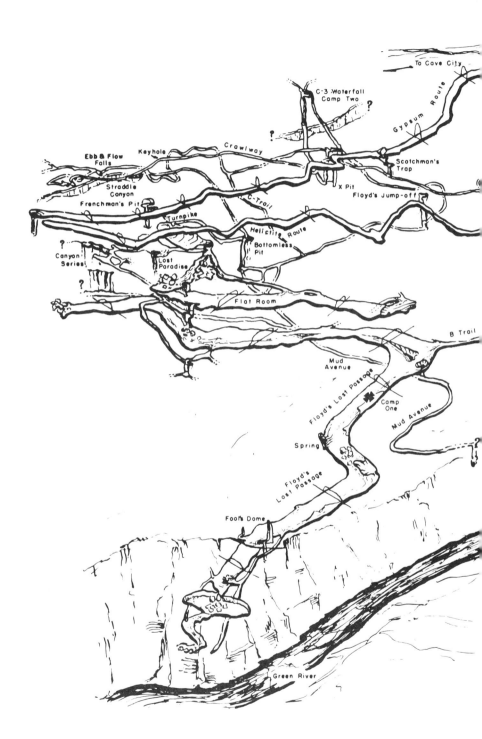

FLOYD COLLINS' CRYSTAL CAVE

37° 12' 50" North, 86° 3' 15" West

A PERSPECTIVE DRAWING OF KNOWN MAJOR PASSAGES
COMPILED FROM SURVEY NOTES OF
THE NATIONAL SPELEOLOGICAL SOCIETY
October 10, 1954
Survey Officer - E. Thierry   Cartographer - John Fisher

Horizontal scale : 1 inch equals approximately 400 feet
Vertical scale   : 1 inch equals approximately 200 feet

That meant Doc had squeezed through the S Curve and was on his way to the Keyhole. I thrust my arms out to turn over, forgetting the walls and banging my elbows. The sleeping bag slid around the first turn, then I inched after it. Halfway around I pushed the bag around the next turn, then again followed it. It was rough going, no doubt about it, but not nearly the torture it had been the first time, or was proving to be on Doc's first trip. Up ahead I could see him trying to squeeze through the Keyhole after Jim, whose light shone through from the other side. Jim was in the parallel passage, the only one that led to the lower levels and unexplored cave.

Doc was trying to jam his head and left shoulder through the Keyhole, but he became wedged. "What do I do now!" he yelled at Jim.

"Try the other shoulder," came the muffled answer. It worked, and Doc pulled his feet through the hole behind him. I stuffed the sleeping bag in and followed.

On the other side, Jim and the Doc were lying down in the passage resting. Sweat streamed from every pore on Doc's face, while the back of his coveralls was drenched; he breathed deeply. "Damn," said Doc. "I don't see how you'd get an injured man through there! You'd almost have to anesthetize him and drag him through." The Doc shook his head slowly from side to side at the tiny hole we had come through. We lighted cigarettes and watched the smoke spiral upward then drift lazily down the passage toward the lower levels ahead.

"You know, Doc," said Jim, "when Floyd Collins came through here, it wasn't nearly so easy. You wouldn't know it, but this Crawlway has been dug out since then. And Floyd used to explore with a kerosene lantern; how he got it through I'll never know."

Doc looked at the wall in disbelief, but sure enough, dirt was piled on every ledge, and a small wooden paddle-like shovel lay on top of one heap. "That's one of the shovels they used, if you can call it that," explained Jim.

We stomped out our cigarettes and buried them in the cool sand floor. In about five minutes we were in a little room I remembered, where three passages led out, not counting the one we had just come through. "Which way, Doc?" asked Jim. Doc screwed up his face and looked at the possibilities. One passage, low and to the left, looked like the best bet. He crawled over to it, noting that there had been some traffic through it, but without saying anything he returned and poked his head up through a high passage leading straight ahead. Then he examined a small crawlway leading to the right.

"I'd say we go to the left," he said. Then he looked at Jim, who was grinning like a Cheshire Cat. True to his prototype, Jim disappeared—up through the hole in the ceiling. The Doc had chosen the wrong passage, as so many would do later on in this three-shell game underground. Doc forced his sleeping bag up through the hole, then followed. When I reached the top Jim was about a hundred feet down the passage—a strange place, with a cross-section like a giant flat-topped toadstool. At the top, the passage was about twenty feet wide and no more than two feet high. Snaking down the center of it was a canyon, perhaps twelve feet deep and twenty inches wide, corresponding to the stem of the toadstool. Red sand covered the floor of the low room, and gypsum facets glinted from the ceiling.

Doc looked at all this, then asked with wonder in his voice.

"How did Jim get down there so fast? Doesn't look like there's room enough to crawl along the edge."

"This is where you do the canyon straddle," I said.

"Sounds like a new dance."

"It'll be very popular this week," I said.

"Hey, Doc!" yelled Jim from down the passage. "Watch this!" Jim placed one hand on either side of the canyon wall and draped his feet down into the canyon. Then he bounded toward us like an express train, doing a kind of leap-frog jump by swinging his feet, bracing them on the canyon sides, moving his arms farther along, and rapidly repeating the move. In virtually nothing flat he was with us.

"Looks like fun," said Doc.

"It is if you watch your footing. One slip and you land on the bottom," said Jim.

Moving along with the sleeping bags proved to be no fun at all, however. We couldn't do a regular canyon straddle and hang onto the roll at the same time, so instead we worked out an alternative by placing our seats on one side of the canyon and our feet on the other. In this manner we could move along sideways, rolling the bags with us on the sand. For about three hundred feet we did this, then left the canyon momentarily and crawled across a low-ceilinged flat place.

We slipped down into the canyon, again carrying our loads over our heads. The top was gone from the toadstool now, and we were moving sideways in the stem. It was a tight squeeze, much like the canyon squeeze in the first part of the Crawlway.

In about ten minutes we reached a muddy breakdown room; the end of the Crawlway, we learned. Large boulders covered the floor, slick with mud. Somewhere above, the sandstone cap had broken, allowing water to trickle in, and with it, mud from the surface. Far below us we could hear water falling among rocks and gurgling along a passage. Even as we sat, drops of water hung momentarily on the ceiling, then plummeted down to spatter on a rock or upon us.

It was now about two-thirty; our trip through the Crawlway had taken us two and a half hours altogether. I still felt fresh; how different from the exhaustion I had felt on the first trip in! Also, I had been able to recognize every passage we had come through. I knew I could find my way out from this point.

"How about some of that fruit?" said Doc.

"You'll be sorry if you eat it all now," said Jim. "It's a long way to the Lost Passage."

Disregarding the warning, I handed the bag to Doc and he ripped off the cellophane. Soon we had eaten half the bag's contents of apricots and prunes.

There appeared to be no way out of this breakdown room in which we found ourselves, except back through the passage by which we had come. Yet we could hear water below. Doc peered around the walls of the room, his feeble carbide light barely reaching its extremities. His face showed the tense curiosity many a cave explorer feels when he encounters a perplexing situation like this one. Would his experience in West Virginia and Pennsylvania caves give him some clue to the way down? He didn't have time to ponder this, for his light went out.

"Better carbide-up," said Jim. We spent the next few minutes recharging our lamps. It was chilly sitting there, for dampness seeps through woolen underwear unless you keep moving. We decided to keep moving.

Jim rose and walked toward one wall; it was just like every other wall in the room, with a pile of jagged rocks near it. When we caught up with him he was looking down a hole barely big enough for a man to squeeze through. Our low angle of vision and the rock pile had hidden it.

"Watch your step going down here," said Jim. "It's about five feet down to the bottom and there are loose rocks here. When you get down, you'll be on a muddy little ledge at the top of a canyon. Don't fall into the canyon. It's thirty feet to the bottom. Same as falling off the roof of a three story house."

Doc gingerly eased himself down the hole. The sound of water was louder now, directly below us. We worked along the top of the canyon, crawling in a half inch of mud toward a place ahead where Jim was starting down, chimneying between the rough, water-cut walls of the canyon. Part way down he stopped.

"Pass those bags down," he said. He took them and tossed them into the darkness ahead, where they landed a fraction of a second later with a hollow thump. Doc climbed down easily, for, after all, he was no novice at cave exploring. But for a while, exploring a big cave like Crystal is like starting all over again. The same bewilderment is there, and the same kind of familiarization process is necessary, only both are more intense than in a smaller strange cave.

We reached Jim and the sleeping bags in the Dining Room, a dome-shaped place about ten feet high and ten feet in diameter with a ceiling open to another passage that followed along on top. One or two rusty tin cans indicated why it was so named. On the walls were arrows, a confusing array. Some were black, having been put there with carbide lamp smoke. Some were scratched in with a sharp stone. They pointed every direction; back the way we had come, straight up, and ahead. Normally, arrows on the wall of a cave passage indicate the way back to the entrance, a standard symbol used by cave explorers for marking their route into a strange cave to prevent becoming lost.

"Jim," I said, "what's the story of all these arrows? I saw them the last time I came in. They don't make sense."

"Well, they do and they don't," he said. "On my first trip in here one night with an old guide, I marked all the arrows pointing in so I could find my way to the Lost Passage again *if* we found it. That was before I knew about making all the arrows pointing out. Then a few other people have put arrows here from time to time, so they'd know this place when they came across it again. Here and there you'll see a dot with a ring around it on the wall. That means that a passage hasn't been explored. Then there are some arrows with a dot above or below them. Those mean to go up or down. Some other arrows have all sorts of lines on them. That's usually the result of someone putting an arrow on pointing the wrong direction, and somebody changing it later."

We walked out of the Dining Room into a narrow crevice lined on each side with grape-like formations, nodules of calcite which had been deposited on the wall when the passage was under water. Some of the projections were sharp enough to tear our sleeping bag covers in several places. The passage was so narrow we couldn't have seen our feet if we had tried. Jim stopped.

"Listen," he said. We could hear him kicking a rock with his foot, then it rolled as though it were going into a hole. Silence. Two seconds later . . . kerplash! It had fallen into water at least fifty feet below. When I stepped over the hole, it felt about six inches in diameter, hardly big enough to put a foot through. We climbed down an abrupt drop off.

"Look out," said Jim. "Keep to the left, whatever you do." To the right a gaping hole led downward into blackness with only the sound of gurgling water floating upward. This was Ebb and Flow Falls.

Doc remarked about the change in the character of the cave at this point. In the commercial portion and through the Crawlway, brown and red sand covered the floor. The walls had been a mellow golden color, covered with gypsum. Now, here at Ebb and Flow Falls, the walls were limestone gray, rather shaly looking and covered with a tan or gray dust. Clearly we were in another rock formation, much lower down than we had yet been. Ebb and Flow Falls consisted of a mere trickle of water coming from the top of a dome at the side of our trail and falling into a pool. When the pool overflowed, the excess ran down into the canyon to join the water we had heard a few moments ago.

Ebb and Flow Falls provides the only drinking water between Scotchman's Trap and Floyd's Lost Passage. *Photo by James Dyer.*

"Anyone want a drink?" asked Jim. I climbed down into the basin and dropped to my hands and knees. The water was cool against my face, and delicious. I drank for a moment, but with trickles of cold water from the top of the dome continuously falling on my neck and back, decided that I really wasn't too thirsty. Doc looked at the water dubiously.

"I don't think we can do that during the expedition," he said. "We'll have too many people running around in the cave and they may contaminate the water. We'll have to give everybody a supply of halazone tablets to purify all drinking water."

"Did you bring any tablets with you this trip?" asked Jim. It turned out he hadn't. And if he had, he probably wouldn't have used them. For one thing, you had to wait a half hour after adding the tablet to the water before the halazone would have any effect. Second, we were all too hot and too thirsty to care. The risks in drinking cave water in Crystal Cave at this time were small, after all, for there were no known sources of surface pollution such as farms or houses. Also, we were at least two hundred feet underground, we thought. Doc removed his hard hat, then buried his face in the pool and swallowed several mouthfuls. He took a deep breath, then drank deeply again.

Jim, veteran of many trips to the lower levels in Crystal, wanted to show us how to drink without removing his hard hat, a real feat if it could be done. He carefully stretched out on the wet stone and lowered his face slowly toward the pool. We held our breaths and watched.

He eventually got his carbide lamp lighted after fishing it and the hard hat out of the very center of the foot-deep pool. "Don't understand it," he muttered.

The passage leading onward from Ebb and Flow Falls was clear and obvious. Again it required canyon straddling or, more properly, canyon hopping. Here the canyon was wider, from two to three feet in width, with more than twenty feet of headroom above. All of us had either spikes or rubber cleats on our shoes, so that by spreading our legs we could literally hop from one side of the canyon to the other with each step. By far the fastest means of locomotion, it was by no means the least dangerous. Below us the canyon yawned downward at least twenty-five feet. Beyond that depth our eyes couldn't penetrate the darkness.

Our trail crossed a vault-shaped room about thirty feet high at the end of which we climbed a pile of breakdown into a higher passage. Its floor was littered with fragments of rock from the size of a fist up to the size of a large watermelon. The going was easy, however, since we could almost stand up. Several hundred feet farther along we stopped to rest. It was four thirty. I knew we were at the brink of the Bottomless Pit, and I wondered how Jim would break the news to Doc.

"Doc," said Jim, "pick up that rock there—that's it. Now throw it over there where I'm shining my light." The Doc obeyed. We watched its graceful trajectory as it sailed out, then started down. I started counting,

"One thousand one, one thousand two, one thousand three, one thousand four." Nearly four seconds later, from far below came a faint but awesome clatter.

"Bottomless Pit?" said Doc.

"Right," said Jim. "Follow me, we're going over." Jim rose and went to the brink of the pit. Along the right-hand wall was a narrow ledge leading to an immense boulder jammed high up against the ceiling. Between it and the wall was a crack about a foot wide at the top and pinching down to nothing at the bottom. Jim crossed the ledge slowly and started head first to pull his body into the crack.

His muffled voice came back to us: "Keep up as high as you can in this crack. If you slip you'll get stuck and may not be able to move." His bobbing light went out of sight. Doc edged along the ledge, entered the crack high, and slipped through.

As I emerged from the crack, I saw Jim's light on the other side, but I didn't see Doc anywhere. My heart raced, but only for a moment. He was sitting on his haunches at the base of the boulder out of my line of sight.

"What now?" he asked me.

"This is the trickiest part," I said. "The floor of this pit is about 90 feet straight down. Ahead of you along this right-hand wall is a ledge only about seven inches wide. Place your back against the right-hand wall, then reach across with your arms to brace yourself against the left-hand wall. There's a steel cable running along that wall. Move sideways, leaning against the far wall for support. If you slip, grab the cable."

"One more thing, Doc," said Jim. "If you fall, you'll die of starvation before you hit bottom."

With these words of encouragement Doc Wanger started to edge cautiously across. Small stones which his boots kicked into the pit pattered faintly on the rocks below. I could hear the scrape of his leather gloves on the wall. He was moving slowly, no hurry. If you have never experienced fear in cave exploring, you experience it the first time you cross the Bottomless Pit. In the dim light I could see the Doc's leg muscles quivering with tension as he braced himself with each step. After what seemed an hour, Doc stuck out his arm, grabbed Jim's hand, and was pulled over to the safety of solid footing. It was my turn.

For me, the complete and utter terror of crossing the pit the first time had gone now and was replaced by a healthy respect for the dangers involved. I passed over both sleeping bags, remembering that all supplies would have to come in this way: every sleeping bag, every item of food, and every member of the expedition.

The ledge Austin and Lehrberger cut in a sheer wall facing a pit considerably shortens the route to Floyd's Lost Passage. The crossing requires the explorer to squirm on his stomach along the brink of the black pit. Every supply team destined for Camp One passed this way. *Photo by Robert Halmi.*

I finally stepped on the solid passage floor which marked the end of the Bottomless Pit. To the left and going down was a passage we had gone through on the first trip. It led down through a maze of crawlways to a fifteen-foot chimney leading to the bottom of the pit. From there we had gone through more crawlways and crevices, and had climbed a dangerous breakdown.

"We're not going the old way," said Jim. "It would be almost impossible to get supplies in that way. Bill Austin and Jack Lehrberger did the support teams a big favor."

Without further explanation Jim led us straight ahead over another canyon-hopping trail, more dangerous than the others because in some places there were no ledges on which to plant our feet. In these short stretches, we relied on meager handholds and footholds to keep us at the top. The passage grew smaller and smaller until it led over a triangular rock at its sharpest point. We had to go over the sharp point of this rock, squeezing our bodies through a tight vertical slit.

"That rock used to bar all progress by this route," said Jim, "until Bill Austin and Jack Lehrberger rounded it off with a sledge hammer. They spent a weekend making that trail improvement and another you'll see in a few minutes."

Jim led us to the edge of a deep pit across which we could see a large room. Here Austin and Lehrberger had shown their ingenuity and persistence by building the "Turnpike." With hammer, chisel and dynamite, they had cut a ledge in the wall just wide enough to snake by the pit. The room itself was immense—elliptical in shape, about a hundred feet long, seventy-five feet wide and twenty-five feet high in the center. As we stepped into it, Doc looked around in amazement, then looked back at the small passage we had just come through. He looked as though he were recounting all the obstacles we'd have to overcome to get an injured man out. The Doc shook his head, whether at the size of the room or the possibilities for evacuating an injured person it was hard to tell. We moved to the left over a floor strewn with rocks the size of sidewalk flagstones. We crawled out through

After hours of crawling, the opportunity to stand erect in Floyd's Lost Passage is a real luxury. The knowledge of the route to this passage died with Floyd Collins and it took Jim Dyer years of searching to rediscover it. The rolling sand floor, the arched ceiling, and the beautiful gypsum flowers found in this passage were adequate reward for Dyer's search. *Photo by James Dyer.*

a narrow passage at the left side of the room and crossed over the top of a big breakdown.

We were in big cave passages now. We walked along one of them, then slid on our rear ends following Jim down a dry flowstone formation. The floor of the passage was bright red now, a kind of sand that marked this part of the cave. Another hole in the floor led us to a broad smooth-floored passage about a hundred feet wide and stretching into the blackness on either side. We crossed this and climbed down another hole into a mud-floored passage. Ahead of us, a crack cut along the floor for about six feet.

Jim stopped and pointed to the crack. "This used to be the only known way down to the Lost Passage, until Luther Miller found another way." He told us to stuff the sleeping bags into the crack, assuring us that we could pick them up down below. The bags fell through the crack out of sight.

We walked down the muddy passage and climbed down into a basin much like that at Ebb and Flow Falls, except that the rock in which it was formed was a brown limestone rather than gray. We stooped through a muddy sewer-like tube leading off on a downward angle and stepped off at the end into a room bigger than anything we had yet seen. Our head-lamps hardly pierced the gloom, and the walls seemed far away. Behind us we heard the drip, drip of water into a pool. A gentle breeze brushed across us momentarily. We were in Floyd's Lost Passage.

The Doc looked tired, yet he was no weakling. Anyone who for the first time had come this far in Crystal had felt the same way, even the most experienced cavers. Now he wandered over to a pile of rocks and picked up one of the tattered sleeping bags from its resting place under the crack in the ceiling. He unrolled it on a flat place in the sand floor and collapsed onto it. We didn't hear from Doc for the next half hour.

Neither Jim nor I was nearly as tired as the Doc. Recognizing where you are and knowing how to cover Crystal's subterrain makes the trip much easier. None of my previous qualms had proved well founded. But I *was* hungry.

Jim and I unpacked the food supply. Cheese sandwiches were unrecognizable as such, long since having been crushed into an unappetizing mass. The peanuts were intact, and we had half a bag of fruit left; not much to eat. We appraised the store of food remaining in Floyd's Kitchen, where Bill Austin and others had left it. One can of corned beef looked promising, and one, without a label, had an "S" stamped on the top—probably for spaghetti. Some instant coffee was there too. Fortunately there was gasoline in a can, so we filled the tiny stove and lighted it; water for coffee was boiling in a jiffy. While Jim opened the cans, I took the aluminum pot and dishes to the spring in a side alcove where I could wash them.

Twenty minutes later we called for Doc to come and get it. He rolled over and groaned something about not wanting any food, but after a moment's consideration apparently thought better of it. The mixture of corned beef and spaghetti was marvelous. I watched the Doctor shovel it down as though it were his last meal. "Good!" he grunted between mouthfuls. Hot coffee made us feel warmer inside. We had soon cleaned up all the food in Floyd's Kitchen and felt we'd never had a better meal—at least underground.

While I washed the dishes again, Jim showed Doc the place where Floyd Collins used to cook his beans when he came here on his solitary exploring trips. The Doctor was intrigued by the few dried beans, now desiccated and preserved, where Floyd had dropped them.

Our original plan called for the expedition's Camp One to be located here at Floyd's Kitchen. That would have been a logical plan if the spring had been running. It wasn't; it was dripping like a leaky faucet, and certainly wouldn't provide enough water for thirty people for a week. Besides, the main supply route, the route we had come in, passed over the spring. Our traverse through the sewer-like tube had already muddied the water.

Floyd's Lost Passage stretched ahead of us like a giant railroad tunnel, or, more nearly, like a New York subway. Doc marveled at the smooth, slightly domed ceiling spanning as much as eighty feet in some spots. Now it was thirty feet over our heads, but it varied from thirty to seven feet as we walked onward. We rounded a turn in the passage at a point where another passage departed from the left wall. Below it lay still another passage, Mud Avenue. Ahead we heard a lot of noise, as though someone had left the water running in a bathtub.

Water poured from a dome in a ceiling niche just off the Lost Passage and plunged into a basin about ten feet below the floor level. We climbed down to it, then looked into the pool. Doc pronounced it an adequate source for water, providing we used halazone tablets.

We retraced our steps to a point where we picked a site for Camp One, a flat place in the passage with a four-foot water-cut trench in the floor at one side. This would make a splendid kitchen with the cooks standing down in the trench and using the main passage floor as a counter top and work space.

We had the information we had come for, so we headed out. We returned to the bend in the passage and Jim led the way down into Mud Avenue, a maneuver that made me realize that we were going back by a different route. The Doc was really going to get "The Tour"! When he came out that night he would feel as hopelessly confused about the cave as others had.

We picked our way down the boulders and finally down a mud slide leading to a trickling stream. Without the sleeping bags, going was considerably easier, enabling us to trot at a fast pace along the muddy stream bank under the Lost Passage. When we reached a special arrow on the wall pointing into a passage, Jim stopped and examined it. "I wonder if this is the one. It's been so long since I was down here last . . ." he mused.

Floyd Collins' bean and kerosene cans lie where he left them in the 1920s in the Lost Passage. *Photo by James Dyer.*

I knew it was the right place from my previous trip, but Jim checked along Mud Avenue for several yards each way while we waited. He said he just wanted to be sure. In we went, crawling now, climbing, squeezing, and walking through dusty passages. We went into the bottom of a pit, not the Bottomless Pit, and looked up. It must have been over a hundred feet to the top. At the far side of the pit we climbed a rubble slope leading to a rock wall about ten feet high. This was Floyd's Jump-Off, a place where Floyd had come one time from the other direction. Floyd had jumped down into the pit on his way in, but when he returned to the spot he couldn't get back up again. He was trapped. Finally he hit upon the idea of piling up rocks until he made a "step" about four feet high. This enabled him to reach some handholds and to pull himself all the way up. We climbed this wall in the same manner, using hand presses and finger holds.

A less ingenious man might have perished trying to find an alternate route out. But in spite of Floyd's prowess as a "one-man expedition," he would be considered reckless by today's cave-exploring standards. Competent explorers always travel with company, except for limited reconnaissance trips away from an exploring group. On these occasions, the explorer invariably sets a time limit on his jaunt and tells his companions where he is going.

The passage ahead continued as far as we could see but, as so often happens in Crystal Cave, we had to take a parallel passage to get on the trail going out. The link to the parallel passage was a small crawlway about six feet up in the narrow passage. In turn, we chimneyed up and slithered through, crawling out from under a ledge along one wall of a considerably larger passage.

"This is the place," said Jim, "where explorers were stopped for many years. You see, Floyd always came in this way, but when he died the secret of getting to the Lost Passage died with him. Look over there." Jim pointed to the mouth of a deep pit to our left. Two obviously rotted ropes dangled into the shaft, hung from some place above. "Mountain climbers came in one time to get down into the pit, but it proved too deep for their ropes. Floyd wasn't a mountain climber, though. He found this small hole we just came through and marked it with a wine bottle sitting in the passage. It's so well hidden that for years nobody found it," explained Jim.

Nothing like this would surprise us now. The only way to explore Crystal, we agreed, was to look behind every rock, look under every ledge, and in the bottom of every pit. All that took time and manpower. We had both, providing no one got hurt, but if someone did, it might take all of us to get him out.

We proceeded through two small rooms and ducked through a tiny hole. "How did we get here?" asked the Doc as he looked around in absolute amazement. He recognized this passage as being one of those we had traveled through shortly after leaving Ebb and Flow Falls. I too had had the same surprise at exactly the same spot. Now I knew the secret of the deception which had been the chief point of confusion for me in the entire cave. "Well, how did we get here?" asked Doc again.

Jim looked at him and smiled, "Right through that little hole, Doc."

Doc still couldn't figure it out. I knew he would be able to after his next trip in. The secret lay in the fact that we had gone up through breakdown to another passage at that point coming in, and on the way back we had come from straight ahead. Lighting from different directions plays strange tricks.

At Ebb and Flow Falls we filled our lamps with carbide and water. I sat on the rubble floor and unscrewed the carbide chamber from the lamp. Carefully I added about twelve lumps of carbide and reached for the top part of the lamp. I knocked against it and it clattered down a foot-wide hole into the canyon below.

"Lost your light, huh?" said Jim. He knew the answer to that question. "Well, I guess you can get out from here with just a flashlight."

"Lend me your light, Jim," I said, "I'm going down after it." I had spotted a place deeper into the cave where it looked wide enough for a man to chimney down to the bottom. I went back to the spot and placed my hands and feet against the walls, inching my way downward. Down on the bottom I found the lamp intact, but rather badly battered. In five minutes I rejoined Doc and Jim. A lamp becomes a personal thing after you've used it to explore dozens of caves, and it isn't given up without expending some effort to retrieve it. At the same time, I hadn't looked forward to traveling the crawlway on my hands and knees nursing a flashlight.

"Charlie Fort dropped one of his gloves in that very same hole one night," said Jim; "took one look at where it had landed and threw the other glove in after it. He said he never did like the way they fit."

We resumed the trip without incident until we arrived at the Canyon Passage just beyond the Breakdown Room at the end of the Crawlway. Doc was working his way along ahead of me when he slipped and fell, banging his knee painfully on the wall. We stopped for a while until the initial pain subsided. What if it had happened down in the Lost Passage? We were happy that we were only about an hour from Scotchman's Trap.

At ten thirty we heard loud voices in the passage ahead, signaling that we were almost at the end of our tour. Jim and Doc went out of the hole and I followed. Flashbulbs popped, and in the glare I could distinguish a crowd of people standing around the brink of Scotchman's Trap. Someone asked me how it felt.

I told him it was a tiring trip, and that we had found a water supply close to a campsite, so we felt quite successful for all our efforts. As a reward for this bit of summing up, someone thrust a bright red apple into my hand.

We snaked our way back along the commercial trail in a blaze of lights. I looked at Doc Wanger carefully; he was limping from the bruised knee, and had the normal Crystal Cave "droop" that comes with a round-trip tour of the lower levels. His coveralls bore the stain of sweat and sand where he had been crawling. He complained of sore knees, something we too felt in spite of our deluxe knee-pads.

"I was going to give physical examinations tonight," he said; "but I just can't. I'm prescribing a good night's sleep for myself."

7

OPERATION BASE CAMP

ON Friday morning, while the Doctor was taking his tour, Bill Austin called Joe Lawrence aside and asked, "Where is the communications section?"

"They haven't arrived yet," said Joe.

"They should be here by now if we're going to have our communications in by Sunday."

"I know," said Joe. "I'm afraid Bob Lutz doesn't know what a big job he's going to have putting wires through the Crawlway to Camp One. I tried to get him here earlier, but he couldn't get away."

"Can't we start laying wires now, at least as far as Scotchman's Trap?" asked Bill.

"No, Lutz has planned in detail just how he's going to put in the communications. We would only confuse things if we tried to start before he gets here."

Just before dark Friday night a panel truck and a station wagon carrying the communications section rolled into camp. Men poured out of the vehicles and began erecting the tent that would be their sleeping quarters. As the crimson sunset rapidly faded into dusk, communications officer Bob Lutz started a gasoline-driven generator and soon had floodlights playing on his busy crew. In the chill of a February evening they went to work.

"How soon can you start laying wire?" Lawrence asked him, since the biggest immediate job ahead was establishing communications.

"We'll start tonight," said Lutz, "just as soon as I get this tent up. But I'm not doing a thing before I get a place fixed to live."

"You have just three linemen in your crew. Can you use more manpower tonight?"

"I'll need more help."

"How about four more people?" Joe asked.

"Good."

Heavy drums of wire were taken out of the truck and station wagon. A switchboard, telephones, and other equipment followed. Lutz put on climbing spikes and lifted his two hundred pounds up a pole. One of his linemen directed the laying of wire from the switchboard to the cave entrance. Another group carried reels of wire into the cave.

Bob Lutz hangs his telephone wires on utility poles so they will be out of the way of the bustling activity of Base Camp. *Photo by Burton Faust.*

When Bob Lutz came down the pole, Joe asked, "How late do you plan to work tonight?"

"We'll work all night and all day until we get communications in."

"Look, Bob, your linemen can't work around the clock without sleep," Joe pointed out. "See that part of your crew gets some sleep tonight so that they can work tomorrow. I can supply you with all the unskilled labor you can use; but if your linemen give out before the job is finished, we're in a real jam."

"Don't worry about my crew, I'll take care of them."

"All right, but remember that it's going to be a terrific job laying wire through that Crawlway. I don't know how you can do it by Sunday morning. But if we don't have communications the expedition will be paralyzed."

"We'll do our best. Can't do more," Bob Lutz answered.

Saturday morning when we walked briskly into camp, Lutz and his linemen were at work. Several telephones in Base Camp were already working, the bells jangling above the hum of activity.

At the switchboard, reels of wire were loaded on a jeep. Bill Austin and Ken Perry took the jeep across country paying out two wires on the ground behind. In this way, they laid three quarters of a mile of wire to the National Park Service fire line that AP Chief Easterly had arranged for the expedition to use. Here, Ken Perry climbed a pole and spliced our wires into the fire line connecting us to the outside world. In case of an emergency the line would be invaluable.

A crew in the commercial portion of the cave laid wire toward Scotchman's Trap. A white dog trotted alongside the jovial group.

"Hey, Doug, I tried to call the supply tent. Operator says to deposit a dime. Where's the slot?"

"Here in my pocket."

Another lineman looked up in amazement. "Look how flat the ceiling is."

"Like it was plastered."

"How'd it get that way?"

"Fellow used to run a still on top of the ridge. Dumped his used mash in the underground river that filled the cave in those days. And . . . well, the ceiling just got plastered."

This was the first team action of the expedition for these men, and their laughter as they toiled showed they were happy that the long planned-for event was under way.

It took two men to handle each of the heavy drums of wire. They would carry the drum on a pipe between them, unrolling wire as they went. Within two hours, several strands of wire were laid over the mile-long route to Scotchman's Trap where two wires were connected to a telephone. Four wires were destined to continue on through the Crawlway.

Bob Lutz's switchboard operator was a stranger to Joe, who became curious to know who he was. He picked up a telephone in Austin's office and said rather directly, "Who are you?"

The operator explained that he was a Fort Knox soldier spending a three-day pass on the expedition switchboard.

"What! You are going to leave us in three days?" Joe asked.

"Yes, sir."

"Wouldn't you like to stay with us for the full week?"

"I sure would."

Well, now, if we had been able to beg and borrow all kinds of equipment and supplies, Joe thought, why couldn't we do the same thing with personnel? A long-distance call to the boy's commanding officer at Fort Knox got him a week's furlough.

Lawrence was pleased with this arrangement until he found the boy's wife was unhappy about it. She wanted her husband to take her home for the week. She took a dim view of caving—even from a switchboard. So it looked as though the expedition was going to lose its crack switchboard operator just after securing him.

Joe hadn't reckoned with the persuasive powers of some of the expedition members. One of our persuaders attempted to change the wife's viewpoint. Not only did he convince the wife that she should allow her husband to work on our switchboard for a week, but he signed the girl up as an extra stenographer for the information officer.

The people assigned to communications weren't the only ones who were busy Saturday morning. Dr. Wanger was now giving physical examinations, and, all day, a steady stream of cave explorers filed through the infirmary to see if they were physically fit for the difficulties that lay ahead.

Saturday noon Base Camp activity reached a higher tempo. As on the day before, Bob Lutz's men descended into the cave laden with wire. The number of reporters and cameramen seemed to treble. Bill Austin briefed the advance party that was to establish Camp One. Doc Wanger continued giving physicals.

Two nurses assisted the expedition physician in these examinations. They gave each explorer a psychological examination consisting of four pages of questions. A typical question asked "Are most people (1) too ambitious, (2) not ambitious enough? Check one." Then there was the step test. Dr. Wanger had each member of the expedition step up and down on a box while he checked pulse rate to determine general stamina. Each explorer had to blow into a peculiar-looking instrument until all the air from his lungs was exhausted. The instrument measured lung capacity.

After Joe had blown into his little instrument, the doctor looked at the dial, shook his head, and handed it back to him. "Try it again," he said.

"Now look here," Joe thought to himself, "you can't flunk the expedition

Ray Streib operates the switchboard in Base Camp while his wife, Barbara, transcribes recorded reports and log entries. *Photo by C. N. Bruce.*

Doctor Wanger thumped, poked, and listened during pre-expedition physicals. Bob Halmi gets checked. *Photo by Robert Halmi.*

leader. I'm taking this examination just to humor you. You can be replaced." Joe blew harder this time, and his face turned red.

The physician looked at the dial and said, "That's good."

We had never seen so many men with notebooks, pencil stubs, and press cameras. They arrived before breakfast and began gathering story material, accepting as fact the opinion of any expedition member, regardless of whether he had been in the cave or not. Although the information officer was available, each reporter was looking for a scoop on what lay below. Television cameramen set up their equipment like itinerant side shows drawing crowds of our personnel who should have been working. Their contribution to the expedition at this time was more confusion. All of the members of the expedition were subjected to a barrage of questions and clicking cameras. High-priority targets for the press were the officers and the women members of the expedition. They were beset at every turn by cameras, tape recorders, and neatly dressed men: "Tell me, miss, what do you expect to find down there?" "I see." "Will you take your lipstick to Camp One?" Always the same questions.

The expedition's own photographers retaliated. Here was an opportunity to get a picture of a news photographer taking a picture. Soon the newsmen found that whenever they opened a shutter it was into the blinding flash of an adversary's bulb.

As interviews followed one another rapidly, consuming more and more of the leader's time, he realized that he would have to cut all the interviews short if he were to get his work done. After curtly refusing to put on a comedy act for television in front of his headquarters tent, he strode over to the infirmary, where Austin was having his blood pressure measured.

"Bill, we've got to do something about all of these cameramen and reporters. They're hampering the expedition's work," protested Joe.

Just then he was tapped on the shoulder. Joe turned around to stare into a movie camera. Its operator said, "Will you look this way and smile, please. We want to get pictures for TV."

Saturday morning Bill Austin had called Roy Charlton into his office to assign him a mission. Bill Austin explained that Roy was to lead a five-man advance party into the cave to establish Camp One at the site selected by Dyer and Wanger. He was then to push on deeper into the cave in search of a site for Camp Two. Roy, one of the expedition's best explorers, would spearhead the assault on Crystal Cave.

Roy assembled his crew and told them their assignment. Roger McClure grinned broadly when he heard he was to be part of the spearhead. This would be the second time he had ventured beyond Scotchman's Trap. Luther Miller looked toward the entrance of this cave he had spent so many hours exploring. A faint smile of satisfaction crossed this veteran's face.

When Roy had completed the briefing of his party, Russ Gurnee strode briskly over to his car to get one of the metal cannisters he and Spence had

The advance party rests at Scotchman's Trap before moving beyond the limits of the telephone line. Left to right: Roy Charlton in the foreground, Ackie Loyd, Luther Miller, and Russ Gurnee. *Photo by Roger McClure.*

built. The first test of the "Gurnee can" was close at hand. Ackie Loyd went to the mess hall to tell his wife, Betty, that he was going into the cave sooner than he had expected. He might not see her again for the next week, for Betty's work as assistant to the supply officer would keep her in Base Camp while Ackie worked hundreds of feet below exploring the complex of Crystal Cave.

The men worked fast putting on cave clothes and assembling their personal gear while in the mess hall Betty helped Alden Snell pack the food Charlton's party would carry. Soon they were ready and Bill Austin bade them goodbye and good luck as they filed off toward the cave entrance. This was the second party the expedition had sent beyond Scotchman's Trap and the first that was to stay below. Bob Halmi of *True* magazine scampered along to get a picture as they went underground.

Roy's party left the bustling activity of Base Camp behind as they passed through the wooden doorway at the cave entrance. They were now five men alone in this vast and peaceful underground world with a definite goal to achieve. It wouldn't be easy. They would establish Camp One and seek a site for Camp Two, but they were experienced men and sixty others stood behind them. They walked briskly down the trail. At Scotchman's Trap Roy reported his position to Base Camp by telephone. This was the end of the telephone line, so Roy would not be heard from again until the communications crew had pushed the line on into Camp One.

Saturday evening when the rush of physical examinations was over in Base Camp, Joe went to the infirmary to have a hangnail checked. It might be infected, so he wanted to do something about it before going underground. Dr. Shoptaugh looked up as he walked in, dangling his finger in front of him.

"I thought I ought to get this looked at before I go in the cave. It must be infected; see how swollen it is?" implored Joe.

"What is?"

"This hangnail. Can't you see it?"

"Oh, that. Hmmm." He squeezed Joe's finger until it turned white.

After a careful examination, Dr. Shoptaugh consulted with Dr. Gehring. Dr. Gehring squeezed Joe's finger until it turned white. The two doctors agreed on a treatment; they prescribed a bandage.

Saturday night expedition personnel assemble in Grand Canyon Avenue for a briefing meeting. Motion-picture cameras are set up to record the event. *Photo by John Spence.*

Dr. Wanger walked into the infirmary at this time and appeared some-what concerned to see the expedition leader undergoing medical treatment. He asked his two assistants what the trouble was. They explained. Dr. Wanger squeezed Joe's finger until it turned white. He agreed with the treatment prescribed, but in addition he prescribed antibiotic ointment to go under the bandage.

"Oh, yes," he said to Lawrence. "You might also try washing your hand."

Saturday the mess hall was a busy place. Ida Sawtelle and her three helpers had served their first meal the night before. This had been after Ida and Alden Snell, the supply officer, had prepared menus, shopped for food, pans, and other kitchen equipment. They were now serving three scheduled meals. Hot coffee was always available, and at night when Ida and her crew went to bed, a night cook went on duty to feed any parties that might come out of the cave before breakfast.

"Brucker, there are some potatoes that need peeling," Ida said to me, ending my pleasant pastime of viewing the swim of activity around me. As I sat in the kitchen cutting out eyes and musing over the way I was spending my vacation, two television news cameramen wandered in. They took shots of my industry, then left.

As we were working, Bob Lutz brought in a telephone and placed it on a window sill. He hooked it to a line another member of his crew had shoved through a hole in the wall. Immediately, the phone rang; Alden Snell was now hooked in. Wasting no time, he discussed tomorrow's menu with Ida.

Ed Easterly wandered in and again wanted to know where the cave lay in relation to the surface of the earth. He had a pair of walkie-talkies out-side and wondered if Phil Harsham, the pool reporter, were to take one handset in, where would be the best place to receive his signal on the sur-face? Having no idea at all, I referred him to Jim Dyer, who scrutinized the landscape through the window, and at last came up with a suggestion. Why didn't Ed stay on the surface and send one of the AP men down into Grand Canyon Avenue with a set? If they could talk to each other, there would be some merit in taking the walkie-talkie through the Crawlway. This answer satisfied Ed, who bustled out to try the scheme. I learned later that it didn't work.

At nine o'clock we finished cleaning the kitchen and went down the hill into the cave to attend the final briefing. Joe Lawrence presided, standing on top of a large boulder in the center of Grand Canyon. Expedition members arranged themselves on nearby breakdown rocks and on the floor, while photographers scampered about on the pile of rocks behind Joe trying to get unusual angles. Joe announced that Charlton's party had already been sent into the lower levels to establish Camp One. Jim Dyer would lead the first party in tomorrow, assisted by Roger Brucker. Those in the party would be Skeets Miller, of NBC, Phil Harsham, pool reporter and staff member of the *Courier-Journal,* and Bob Halmi, photographer on assignment from *True* magazine.

Bill Austin reminded the party to be careful of formations, that they were fragile and could easily be broken by careless exploring. He gave several suggestions regarding sanitary considerations, and reminded each person to carry a supply of "lens tissue," there being no convenient leaf-bearing trees in the cave.

I wandered over to Jim Dyer who was talking with Miller, Harsham, and Halmi. "Roger, I want you to meet Skeets Miller," said Jim.

"How do you do, Mr. Miller," I said as he extended his hand.

"Call me Skeets," he said, "I have a feeling we're going to get to know each other pretty well these next few days." His face wrinkled up in a friendly smile you couldn't help liking. On his head was a shiny black hard hat and a brand-new carbide lamp. He wore a blue flannel jacket with the NBC insignia on the pocket, and a pair of black wool pants which he had tucked into a pair of new high rubber boots. It occurred to me that he would fit to a "T" the Hollywood idea of a cave explorer, but he couldn't be blamed for that. It had been a long time since he had been in a cave— Sand Cave—that awful death trap where Floyd Collins met his end. I wondered how he had managed to crawl where more powerful men were fearful to go, when he brought food to Collins in those last days. Few remembered his heroism now, remembering him instead for his Pulitzer Prize-winning stories filed from Sand Cave during those terrible days of waiting. But Crystal was a much longer cave, where oftentimes you crawled for hours. It took nerves, yes, but it also took a great deal of stamina. Would he be able to make it? I asked myself.

Phil Harsham of the Louisville *Courier-Journal* wore a white hard hat and brand new coveralls. He was a large man, over six feet tall, with a jutting jaw. He looked as though he could take care of himself without trouble. He spoke with a gentle voice and with an eager sincerity, wanting to know what he should take into the cave with him tomorrow.

Bob Halmi was unlike the other two; he had a slim, short frame and eyes that twinkled with good-natured curiosity. He wore a red-checkered shirt and on his head a glistening hard hat perched at a jaunty angle. Two Leica cameras dangled from straps around his neck, a Rolleiflex was cradled in his fingers. He was well on the way to a heavy beard. "Hi, Brucker," he said in a foreign accent I judged to have originated in the Balkans. I went through the inventory for the third time.

Small groups chatting informally began to break up. Ahead of me, Joe Lawrence climbed the trail out of Grand Canyon to the entrance. He walked slowly, his head and shoulders drooping like a man with many cares.

"You look tired, Joe," said Doc Wanger.

"I am . . ." he said, with a look betraying the fact that his continuous activity was too compelling. His mind could not retreat to rest.

Some of us had seen the signs of pressure in Joe's eyes. It had been this way for three days now as the many activities of organizing the expedition had gone on simultaneously around the clock. The night before he had lain down on the floor in Bill Austin's living room long past midnight to toss the few remaining hours before daybreak. The previous night had been the same—a few restless hours in a sleeping bag in the cave a hundred feet from Floyd's casket.

He had told me he feared tonight would be the same. He desperately needed sleep; but an endless string of thoughts and problems tramped through his brain. Would communications be established in time? Could we move in supplies fast enough? What had he forgotten? What essential item had been overlooked? Would these people unfamiliar with this confusing cave get lost? Would the whole expedition flop on its face—a complete farce?

"Can I do anything for you?" Doc asked Joe when they reached the crest of the hill.

"I hope so."

"Trouble sleeping?"

"Yes, could you give me something to knock me out?"

"There's a spare bed in my room at the Mammoth Cave Hotel. Sleep there tonight. When you go to bed, take a couple of these sleeping pills."

"Thanks, Doc."

At the hotel, Joe took a shower, crawled into bed, and took the prescribed dose of sleeping pills. He lay sleepless for two hours planning the explorations he would direct from Camp One. Then Dr. Wanger arrived and told him to repeat the dose of sleeping pills, which he did. Doc returned to Base Camp for an hour and a half. When he came back to the hotel, Joe was still awake. Doc gave him a couple of capsules to swallow. They talked briefly of the trip they would make together on Monday to Camp One, but soon, only Doc's voice filled the room. He turned to look at Joe, now fast asleep.

When Joe awoke at 10:00 A.M. he felt refreshed. A big breakfast in the hotel restaurant was the final tonic he needed to prepare him for seven days beneath Flint Ridge.

The expedition had a firm grip on the entrance. Base Camp was a smoothly functioning facility. A flood of supplies and personnel was ready to cascade into the cavern. The immediate objective would be a build-up of strength at Camp One six hours from the cave entrance. We must move in food, sleeping bags, mess gear, and scientific instruments. We must send in explorers, surveyors, cooks, and scientists. We would develop more strength at Camp One than had ever been concentrated in any cave before.

8

ADVANCE PARTY

A<small>T FOUR IN THE AFTERNOON</small> on Saturday, five hours before the personnel briefing in Grand Canyon, twelve husky men lined up in front of the communications tent. Bob Lutz looked them over, knowing that on these men rested the success of the expedition. Lutz was direct and to the point.

"Your job is to push four little phone wires into a mighty big cave. Some of you won't be getting much sleep for a long time. You'll be tired. But let's see if we can get a phone down there tonight. Charleton's advance party is counting on it." For all practical purposes, the advance party was cut off from the world. Their findings and their needs could not be determined without telephone communication.

The men hoisted spools of wire to their shoulders. Some carried field telephones and others had transformers stuffed into coverall pockets. They had eaten a big meal two hours earlier since there would be no hot food in the Crawlway. Phil Smith, Jack Lehrberger, and George Welsh, who knew the Crawlway, would choose the route the wires would follow to minimize accidental damage to the circuits by subsequent travelers. Henry Douglas, a lineman, was in charge.

In Grand Canyon the procession met Bill Stephenson's crew placing previously-laid lines off the trails. A wire spool dropped from the arms of one member, rolled down into Grand Canyon, and tore out a floodlight fixture. Bill Austin muttered, "If you have trouble here, what's going to happen when you get into a passage with no room overhead?"

At the Trap, the party stopped. George Welsh turned to Phil Smith. "I'm for getting a couple of spools and taking off." The last words they heard were those of Douglas reporting in. "Base, this is the Trap . . ." They climbed down into the Crawlway and removed the binding wire from the spools. Two men led off unwinding the bulky drums; a third followed hanging the wires on overhead stone projections out of the way of explorers.

A caver lays wire in the Crawlway by rolling a spool ahead of him.
Photo by John Spence.

The narrow slit was pinching down now. The party rested at Last Chance.

Pushing the spools ahead of them was no easy task. Time and again they pushed and the spools wouldn't move. Sand was piling up in front of the roll, requiring bare-handed shoveling to clear the way. The towel Welsh wore around his neck was already wringing wet. After each ten minutes the men changed jobs, so each in turn could bask in the comparative luxury of hanging wires out of the path. In the low part of the Crawlway they devised ingenious methods for keeping the wire out of the way. One method was to place wires on ledges, weighting them down with rocks; another was to pull the wire through unused cut-arounds; a third, to drive spikes into wall crevices and hang wires on these. The group wondered why the others had not caught up with them by this time, since the work was going slowly. The tight part of the Crawlway was no place to wait for anyone, so they continued onward.

Their arms ached and the spools seemed to get heavier with each foot of wire unreeled. At the Keyhole they decided that putting the wire through it would be an invitation to circuit trouble each time a support group passed through. They discovered that the spools might pass through another narrow slit, by-passing the Keyhole, if rocks were cleared away to enlarge the route.

The sight of these wires entering an impassable place did not deter some of the early support teams as yet unwise to the way of the cave. They had orders to follow the phone lines, and follow them they did, even to the point of moving a large pile of rocks to enlarge the route. The Keyhole By-Pass, far more difficult than squeezing through the Keyhole itself, was the chief reason why a few members of the support group were unaware of the existence of the Keyhole at the end of their first round trip.

At Straddle Canyon Smith wrote a note for wire parties following, telling them that wire was to be laid in the bottom of the canyon where it would be out of the way of support parties traveling at the top. Near the end of the Crawlway, Lehrberger surveyed the walls of the passage. "There ought to be a short cut through here somewhere. I'll go around the usual way and yell. If you hear me, then we'll try to shove the wire through that breakdown over there."

In a few minutes Smith and Welsh heard Lehrberger calling: "Looks O.K. See if you can pass them through." One spool went through the tiny crevice; the other was empty, so they waited for the others to catch up with them.

Waiting in Straddle Canyon was about the most uncomfortable thing imaginable. When they dangled their legs into the canyon they had to hunch over to avoid the low ceiling. When they lay out flat on the damp sand they became cold. Even the crackers and raisins in their lunches failed to raise their spirits.

"I can't stand this waiting," said Lehrberger. "I'm going back to see what the hold-up is." Neither Smith nor Welsh felt equal to following him.

One reason for the delay, unknown to Smith and Welsh, was that the expedition had suffered its first casualty. Shortly after one of the last wire-laying parties had entered the Trap, one member, pushing a spool of wire ahead of him, cried out in pain. He gasped that he thought he had broken a rib. The party spliced their phone into the line and reported to Base Camp. While the infirmary prepared for the emergency, Ken Perry started out of the cave aiding the man as best he could. Later examination indicated that the injury was a bruise rather than a break, but the man was limited to the commercial routes for the remainder of the expedition.

Inside of fifteen minutes, two marines on leave for the expedition hove into view. They carried two 25-pound cans of carbide, the precious lumps of acetylene-producing fuel for explorers' headlamps. "Hello," said one. "This stuff is probably powder by now." The marines told the waiting pair that the wiring party was still going strong, but that the Crawlway was giving them a rougher time than any of them had experienced in a cave before.

When the parties met, fresh rolls of wire were spliced on. Traversing the muddy ledges near the Dining Room forced them to pass their loads forward from man to man. Short cuts in caves save distance, but in Crystal they didn't always save time. Since no one knew how far it was in feet to the Lost Passage, every effort was made to conserve wire, even though Bob Lutz had his truck stacked full of spools. Spools of wire in a truck had little in common with spools of wire at the far end of the Crawlway. The labor to move them there made the difference.

They passed over the top of Ebb and Flow Falls through a murderously tight crawlway floored by jagged rocks. No one had ever used this route, all the more reason why wires would be safe. Loops of wire slipped sideways off the coils, making handling even more difficult.

Easier passages led to the Bottomless Pit. Here the group paused to gape over the brink of the first big pit they had seen. One member of the party described his impression of it this way: "I had a three-cell flashlight and I pointed it at the bottom. Somewhere along the way it just ran out of shine."

Shortly after midnight Bill Austin strode into the communications tent impatiently seeking news on the wiring operations. The operator raised his fingers to his lips, then pointed to the tape recorder revolving slowly. A three-way conversation was going on between communications party leader Henry Douglas down in the cave, switchboard operator Ray Streib, and communications officer Bob Lutz.

STREIB: All right. Yeah. Yeah. Who is this?

DOUGLAS: This is Douglas at the top of the Bottomless Pit.

STREIB: At the top of the Bottomless Pit?

DOUGLAS: We have strung wire this far. We are going to the site of Camp One. We hope within the next hour.

STREIB: Within the hour? You'll be down there all night?

DOUGLAS: No. We will rest a while and come back.

STREIB: Uh-huh.

DOUGLAS: We are not prepared to stay all night.

STREIB: Bob wanted to talk to you when you called in.

DOUGLAS: Is he there?

STREIB: So if you will hold on just a minute I'll connect you.

DOUGLAS: All right.

STREIB: I am going to connect you to the supply phone.

DOUGLAS: O.K.

STREIB: O.K. Go ahead.

DOUGLAS: Hello.

LUTZ: Yoo.

DOUGLAS: Bob?

LUTZ: Yeah.

DOUGLAS: What's new?

LUTZ: Where are you?

DOUGLAS: At the Bottomless Pit.

LUTZ: You are at the Bottomless Pit?

DOUGLAS: Yeah.

LUTZ: How is it going?

DOUGLAS: Going fine. We have run out of wire for a third line. We have the long wires which extend into the Lost Passage.

LUTZ: Yeah.

DOUGLAS: Then if we have some left we will bring it back and try to make it a complete four wires. If not, somebody later will have to bring another roll of wire and complete the connections.

LUTZ: How is your crew holding up?

DOUGLAS: We have two or three who are getting sort of pooped out and they are going to wait near the Bottomless Pit for us to come back.

LUTZ: Uh-huh. Well now, I'll give you a clue. They have a hell of a forest fire going up here.

DOUGLAS: Forest fire?

LUTZ: Yeah.

DOUGLAS: Where?

LUTZ: Well, outside the camp here a mile or three, and they are gathering up everybody that can walk. So for right now don't be in any hurry about getting out. But if any of the boys are getting kinda burned-down, why, start them out. And do as much as you can without completely burning up the works there now, that is, without wearing the boys clear out.

DOUGLAS: You are not going to send in any party tonight?

LUTZ: They took 'em all.

DOUGLAS: O.K. We will finish up what we can do. Then move out gradually.

LUTZ: And hook your phone on the line in.

DOUGLAS: O.K.

LUTZ: While you are there, try out the other line.
DOUGLAS: I can't. We are not back to the end of that line.
LUTZ: You are not going that far then?
DOUGLAS: Yes. See you before daylight.
LUTZ: O.K.

The seriousness of the forest fire had been brought home to the expedition when tires ground to a stop outside the communications tent. A door slammed and those inside went out. A tall, stout man in the green uniform of a Park Ranger addressed them.

"Got two forest fires," he said, "both out of control." The sky was a dull orange to the northeast.

"Wonder if you've got any volunteers?" he asked.

Bill Austin sighed and looked toward the blaze. "How many can you use?" he asked the ranger.

"About a dozen will do fine, if you can spare them." Spare them? There was no question about it. The fires had to be put out at any cost, or thousands of acres of timber would go up in smoke. The stiff breeze now blowing was driving the walls of flames deeper into the tinder-dry forest. If not checked, the fires might devour the National Park, Floyd Collins' Crystal Cave property, and perhaps even the expedition itself. All thought of the immediate needs of the expedition were set aside. Bill Austin had his first major headache.

About fifteen minutes later a tired group of men assembled at the office. Their names were written down and each was issued an electric headlamp and hard hat. Several were given back-pumps to operate. Another ranger had arrived with a truck by this time, so the group split into two parties. Within two minutes the entire group had vanished down the road, leaving Crystal Cave with a vastly undermanned expedition. How long would the fire last? In what shape would the already tired men be when they returned? And what about the men below who were depending on uninterrupted supply and communications? An accident down in the cave now would spell disaster for the expedition. Bill appeared to be keenly aware of these questions. He bit his fingernail and looked after the recently departed crews.

A haggard wire party deep within Crystal Cave pushed on across the Bottomless Pit with copper life lines to Camp One. Some of the men quivered at the sight of the pit. They excused themselves from going across. The two marines with Smith and Lehrberger passed spools along the narrow ledge.

"There goes the wire!" One spool shot toward the bottom. Seconds later it hit with a thud. There was only one comment, ". . . It *is* a long way down."

The others, now rested, rejoined the party, bringing the strength up to thirteen men. One still refused to cross the pit. They pulled on the taut wire running into the abyss, unwinding the spool while a third man wadded up the slack in a tangled mass. They were under way again.

As the number of spools diminished, so did the enthusiasm of the group, in spite of the walking passages they were in now. They came to the end of their wire at the bottom of a flowstone slide still some distance from the Lost Passage, but the kind of distance a cave man likes: head room, not much crawling, and a fast route. They knew the advance party could contact Base Camp in a matter of ten minutes if the need arose. Lehrberger suggested that some of them check Camp One to determine their supply situation and tell them of the location of the telephone. Smith volunteered to go with Lehrberger, the others preferring to rest for the trip out.

They returned with the news that everyone at Camp One was sleeping so soundly it would be a shame to waken them. Douglas twirled the crank. An alert operator had been keeping the circuit open in case of trouble on

Support personnel who should have been resting Saturday night had to don electric headlamps and go out into the night to fight a forest fire. *Photo by John Spence.*

Crossing the Bottomless Pit never became a routine. After working across a wide ledge (shown) the explorer comes to a six-inch ledge (not shown) where nerve and a steel cable keep him from falling into the 90-foot depths. Left to right: Bill Wanger, Charles Kacsur. *Photo by Charles Kacsur.*

the lines, since by this time he fully realized the difficulties of setting up a complicated phone network. The ringing mechanism was out of order, but a cheery voice on the other end began to speak.

The phone line would be monitored in the morning, they were told. Douglas suggested that one of them leave a note to this effect for members of the advance party. Smith returned to the Lost Passage carrying a can of carbide and the note:

> Hello you cave dwellers. You will find the phone near the flowstone. Supply men are all out fighting forest fires, so don't expect anyone in until late in the day. The ring on the phone is not working. Surface will monitor every hour on the hour starting at 7 A.M.

This was not a very happy "good morning" for Roger McClure, who would find it. The party headed out.

At five thirty in the morning on Sunday the fire-fighters arrived back in camp. In their exhausted condition, they would see no service for at least twelve hours, for they had helped bring the fire under control and had worn themselves out in the process.

Joe Lawrence made first contact with the advance party deep within the cave. Because of the significance of the first report it was tape-recorded so staff officers could pick up useful information and hear for themselves how the expedition was progressing.

LAWRENCE: Hello, Lawrence speaking.

McCLURE: This is McClure in Floyd's Passage.

LAWRENCE: Yeah.

McCLURE: We arrived in here at 8:00 last night.

LAWRENCE: Just a minute. (Aside) Are you recording this?

McCLURE: It took us five hours to get in. We were lucky to get to bed at a decent hour and got nine hours sleep. Sent three of the boys out on B Trail to check for location for a second camp.

LAWRENCE: Who were those three?

McCLURE: They are Roy Charlton, Russ Gurnee, and Ackie Loyd. Luther Miller and myself stayed at Floyd's Passage. Located a site for a dining room, and are checking on a site for the latrine. At the present Luther is doing a back-track of the telephone line, and to pick up another party coming in to give them a lift. Did Phil Smith talk to you?

LAWRENCE: No, he didn't.

McCLURE: Well he arrived in about one o'clock last night. With one communications line. Those white sea bags, those navy bags?

LAWRENCE: Yes?

McCLURE: Those packed full are about the largest parcel we can get through. They worked out pretty good.

LAWRENCE: O.K. We will not send in anything larger.

McCLURE: Now you can pack them full. We stuffed them. But take small rope and wrap them. In other words, the handle on them will not stay fastened.

LAWRENCE: Can you hold the line a little? I will get a phone somewhere else. I have the switchboard tied up right now. O.K., go ahead with your report.

McCLURE: Those navy sea bags work out fine. The handles tear very easily so the thing to do is to get some small cord and tie them tightly in two places around them. Then join them like a long handle.

LAWRENCE: Will you repeat? I am having trouble reading you.

McCLURE: You should bind up everything that comes in, with small cord of some type, sleeping bags, sea bags, any parcel, should be wound or lashed well, with some type of small cord, in a way that it can be handled. With some sort of handle on it, and also a dragline.

LAWRENCE: Stephenson says he will take care of that with all parcels going in now. They will be sent in that way.

McCLURE: Do you know what supplies are coming down with the party starting in this morning?

LAWRENCE: Just a second. I have a list here in my hand. You'll have condiments and miscellaneous supplies, for the kitchen, gasoline, Jim Dyer is bringing in a party of Halmi, Miller, Harsham, and Brucker. They will have candles and a gallon of gasoline. We have another support party leaving an hour behind them. They are led by Tom Barr, and the group consists of Richter, Shoptaugh, and Gehring. They will bring in condiments, the bags for carbide, a stove, a gallon of gasoline, three personal kits for the reporters (those personal kits also contain liners for the sleeping bags), repair kits for carbide lamps, food, supper for tonight, breakfast, lunch and supper for tomorrow. All meals packed for ten people. The ten are the five that went in with you plus the five that are coming with Jim Dyer.

McCLURE: What about the other party?

LAWRENCE: Barr's party is a support party and will return to the surface immediately after arriving.

McCLURE: Is the support party to be fed down here?

LAWRENCE: Barr's party is bringing their own lunch so you will not have to feed them.

McCLURE: Fine.

LAWRENCE: We also have working in the cave Bob Handley and Hugh Stout, trying to wind up the communications system.

McCLURE: You might be interested in the load we brought in. We figured that it would be about the maximum load anyone would want to bring in. Now there were five of us. We had three sleeping bags, one navy duffel bag, one Gurnee can, three small knapsacks.

LAWRENCE: What was in the knapsacks?

McCLURE: Some personal gear. The sea bag was also full of personal gear.

LAWRENCE: Were you able to pack five personal bags in a sea bag?

McCLURE: No.

LAWRENCE: How many were you able to pack?

McCLURE: Three, maybe four. In fact none of us actually has what was stated in the original documents. That was too much to take along. There are three of us that have most of it. And that is just about all the sea bags will take.

LAWRENCE: That is stuff for three people?

McCLURE: Yes.

LAWRENCE: Well, I will remember that and see that we pack them that way in the future. How is the physical condition of all the people?

McCLURE: Fine, very good. We got a good night's sleep and everybody is fresh as a daisy.

LAWRENCE: Let me check the times on you. You said you arrived at Camp One at 8:00 P.M. last night. What time did Charlton leave this morning?

McCLURE: At 10:00. They got a late start, inasmuch as we found a note at the phone that the surface party was out on a fire. What was that, a forest fire?

LAWRENCE: Yes, we had to dispatch twelve of our people to fight the fire last night. So we are behind here on the surface.

McCLURE: We held up until we could make contact with the surface. I suppose it was about 9:30. Then we got them started off about 10:00. They are to be back at about 6:00 tonight. Possibly later. We are giving them some hours later.

LAWRENCE: You expect them to return about 6:00 P.M.?

McCLURE: About 6:00 to 9:00 is the way we set it.

LAWRENCE: This afternoon we will send in the two television boys also. Here are some instructions for you. Tell Jim Dyer that he is acting leader of the exploration group. He will hold that position until Austin arrives sometime later tonight. Tom Barr, leader of a support party, is bringing in a log. I want Jim Dyer to open that log and start making entries immediately of the activities of the exploration group.

McCLURE: Jim Dyer is privileged to go out on exploration?

LAWRENCE: That is right. Jim Dyer is unlimited. He will be in charge of the exploration group. You had better tell him not to go out on exploration though until after Austin arrives. I would like for him to stay in camp and take charge. He can go out on short runs for a few hours, if he would want to help the photographers get pictures. If he wants to do any more have him call the surface. The switchboard has asked me to set a time at which we can contact you again. You have a watch?

McCLURE: Yes.

LAWRENCE: What time do you want to call back here?

McCLURE: It is immaterial to me inasmuch as I can make it any time.

LAWRENCE: Stand by while the recorder is reloaded. (Later) O.K., the recorder is running again. Stand by while I check with the switchboard operator when he would like you to call him again. Give us a call at 2:00.

McCLURE: 2:00?

LAWRENCE: Yes. If Dyer arrives have him call me. Or say, let us leave it this way: When Dyer arrives regardless of how late it is have him call me.

McCLURE: Well, I tell you, I don't believe we can ring out, if the crew doesn't get through.

LAWRENCE: Oh, I see. Hold on just a minute. If I know within two hours of when he arrives that will be good enough; at 2:00 the switchboard operator will make contact. In case you have to get a call through at any time, before we get this ringing circuit running, you can get him on the hour at any time. He says he will cut in on your circuit and monitor it, for five minutes following the hour.

McCLURE: Fine. O.K.

LAWRENCE: And we will hear from you again at 2:00?

McCLURE: Yes, O.K.

Camp One had been established on schedule. Charlton, Loyd, and Gurnee departed to spend the day in search of a site for a second camp on the edge of the endurance barrier. They had been instructed to concentrate their efforts in the B Trail–Bogardus Waterfall Trail area, two little known avenues leading into the unknown. While they were gone, McClure busied himself by putting the camp in order, and looking into a few obvious side leads off the Lost Passage.

A bitter, disappointed reconnaissance party returned to Camp One. They had penetrated for three hours into B Trail, to a place where the floor was covered with four inches of mud and water, with only a twelve-inch space between the ceiling and floor. They had found no suitable site for a camp. However, they had found a number of inviting side passages which time had forced them to abandon. Prospects for establishing a second camp early in the expedition faded.

9

PERSONNEL BUILD-UP

IT DIDN'T SEEM LIKE SUNDAY when I awoke, but it was. Through holes in fleecy ragged clouds, patches of brilliant blue popped through. Sunshine streamed down on our camp at intervals, playing its warm yellow on billowing canvas tent roofs. Then it would grow darker, almost cold as a cloud blocked the light. Men in winter jackets, with hands thrust into pockets, scurried with pliers, clip boards, and rope from one tent to another. The sun came out again and men walked more leisurely, sniffing the brisk air and stopping to look at the bluff on the far side of the Green River.

Sunday, February 14th, the official kick-off of the expedition. I hurried over to the mess hall and ate breakfast, knowing we were scheduled to leave with the press party at eight thirty. There I learned that lunches were being prepared for our party to eat underground and saw Ida putting peanuts, dried fruit, and candy bars in a plastic sack. Would that hold us until the next supply team arrived? I didn't think so. The people scheduled to go in on support for Camp One after we departed had just returned dog-tired from fighting forest fires. They were exhausted, certainly not up to a trip to the lower levels carrying supplies.

I went to my station wagon, where I donned a pair of coveralls and laced on my caving shoes. Already the toes were showing signs of wear, yellow scuff marks on the brown leather from just one trip through Crystal. They had seen use before, but nothing like the beating they were getting from the rocks and passage floors of Crystal Cave. I cleaned out the white spent carbide powder from my lamp, a sure indication of my weariness after the last trip to the lower levels. Normally I empty the lamp immediately after coming out of a cave, but this time I had simply forgotten about it. I filled to the brim my extra carbide box and water bottle. Into my rucksack went my sweatshirt, several cans of surplus rations, and the lunches. The sack wasn't full by any means because I knew I'd have to save space for gear that would not fit in the pockets of the press party. They had been warned to travel light, but novices to Crystal invariably try to carry too much gear along; it is usually abandoned along the route to the Lost Passage. With new batteries in the flashlight, I was ready to go.

Jim Dyer, Phil Harsham, Skeets Miller, Bob Halmi, and Ed Easterly stood in a tight knot in front of the wind-whipped communications tent. Phil asked me: "What's the best way to carry this five-cell flashlight?"

"You'd better leave that," I said. "Otherwise you'll throw it away when you get into the cave. There's absolutely no comfortable way to carry a five-cell flashlight when you're crawling." He handed the light to a friend and promptly drew out a two-cell model from his pocket! True to form, he was planning to travel heavily loaded. Slung around his shoulders were two rucksacks, similar to mine but jammed full of "essential" items, spare clothes, shoes, and other gear. Skeets Miller had a pair of sacks slung around his neck, also stuffed with "essentials". Beside him on the ground rested a large fishing-tackle box.

"Does this go?" I asked as a joke.

"Yes," said he, to my astonishment. "It's the amplifier for my broadcasts." I picked it up. It must have weighed thirty-five or forty pounds.

Skeets Miller, who was clean, shiny, and enthusiastic, carried, among other things, a heavy steel box containing an amplifier for his radio broadcasts. *Photo by Robert Halmi.*

Halmi carried his two Leica cameras in front of him. I wondered if they wouldn't get smashed in the Crawlway. "No," he said, "I've fixed them so they fit inside my coveralls, like this." He unbuttoned his chest buttons and neatly slipped both cameras in. They still bulged, however, and I could see that even this arrangement would cause him some discomfort. He also carried a rucksack, stuffed full.

Bob Halmi's new sea-bag of photographic equipment caused him to list a bit to starboard as he walked. *Photo by Robert Halmi.*

"What's in this?" I asked, pointing to a white canvas sea bag. After the amplifier was officially listed as part of our load, I had no doubt that the sea bag would go with us too.

"Electronic flash," he said, "and camera gear—enough film for 3,500 pictures, both color and black-and-white." I lifted the sea bag and groaned. It weighed at least fifty pounds, perhaps more, and was filled with delicate electronic flash and camera gear! I wondered what he would do with the pieces when and if the bag arrived intact in the Lost Passage.

A photographer asked us to line up for a group shot. Afterward I whispered to Jim, "I thought you said they were traveling light!"

"You should see the stuff we made them leave behind!" Dyer looked at his watch and said it was time to get ready. I took a few items from Phil Harsham and secured them in the rucksack. Phil confessed to me that he was a bit apprehensive about the whole business. He hadn't had much sleep last night, and he had risen at six to get ready. All he had to eat was a bowl of cereal and a glass of orange juice. I too became apprehensive at this point. I was aware he didn't know the full story about the support teams and how they might not be able to arrive on schedule. It might mean that he would go hungry for the first twenty-four hours, at least. But we were walking down the hill into the cave now.

In the bottom of Grand Canyon we stopped for what I supposed was some last-minute picture taking. Our party lined up in the glare of the motion-picture lights where we were told to take off our hats on signal. Suddenly, out of the gloom, a strange voice began reading from the Bible. I couldn't have been more startled if it had been Floyd Collins himself.

I learned later that the man speaking was a local minister who had read from Psalm 19 as a memorial to Floyd Collins and as a beginning note for the expedition. "The heavens declare the glory of God; and the firmament sheweth His handywork. . . ." he read. In the fourteenth chapter of the book of John, fifth verse, is a thought that might have been more appropriate: "How can we know the way?"

Carl Spalding took the sea bag from me and offered to carry it back to the Trap. Someone had already taken the amplifier from Jim. I protested momentarily, but thought better of it, because I'd have plenty of time to carry that sea bag where the going was much more difficult than here in the commercial route. The idea of men volunteering for Sherpa duty was a fine one; I would remember it on the next expedition. All of us walked along munching apples, a custom now firmly established.

At the Trap we stopped to put on knee-pads.

"Is this really as rough as they say?" asked Halmi in a low voice.

"Depends on what you're used to," I said. "Personally, I think it's pretty rough." I had heard stories from the others that Halmi was used to plenty. He had come to this job from covering a ballooning story for *Argosy*. Prior to that he had paddled up into the frozen north in a kayak for the *National Geographic*. What could we learn from this adventuring man? What would he learn from us?

A flashbulb popped as we posed for the one last shot, then the inevitable "one more."

Bill Stephenson was on the phone arranging with the information people to tape-record a running account of our triumphal entry into the Trap: "Jim Dyer is ready to start in. He is carrying a large box belonging to Skeets Miller. He has a nylon safety rope wrapped around him and looks in the pink of condition. He has just rested and had a smoke. Jim Dyer is now going down into the Trap. Ida Sawtelle, as part of the support team, is passing equipment down. Skeets Miller is now getting ready to go. He's shouldered his bag and started down through the Trap. Bob Halmi is following Skeets. He also has a safety rope with him. It is his own personal equipment. He's loaded down completely. All the men have their knee-pads adjusted. Harsham has now adjusted his equipment. He's making sure he has plenty of water in his canteen. He has taken off his sweater and has that attached to his belt outside so he can put it on when he gets through the crawl. Brucker is now passing the large sea bag of equipment they are taking. That is being lowered into the Trap. Moving-picture cameras are grinding away. We have floodlights at the Trap and are making a photographic record of this. The last of the equipment has now been passed down. They are posing for a few shots. There are only two men left outside the Trap. There are more reporters and photographers outside than in."

"Do they seem very nervous about the whole thing?"

"No, they seem very calm. They don't know what they are up against. Do you hear the background noise?"

"Yes, it comes in very clear."

"They are passing instructions back. Each man reported them back to the next. Ida Sawtelle is now coming out of the pit. There is nobody there now but the official party. The last picture is now being taken. Norling of CBS is taking pictures. One photographer gets irritated at another photographer. There goes Halmi down and Brucker is following down in back of Halmi. The passage is practically empty, all of the equipment having been passed forward. He has knee-pads, and a flashlight hanging from his safety rope, and his own personal pack. You can see him now just disappearing. We can see the red dot on his hat reflecting in the lights as he passes beyond Scotchman's Trap. There he goes."

"Is that all?"

"Yes. Last man in. The supporting party is now returning to the surface."

We left the buzz and hum behind us—we were on our way. The sea bag seemed to weigh a ton! I made a feeble pretense at being careful with it, knowing that its contents cost several hundred dollars, but I realized that it was just too heavy to worry about. The boxes inside clunked and thumped as I dragged the bag over the sand floor, then pushed it down into the narrow canyon after Phil, who was making slow progress through the tight squeeze. His rucksacks caught and held, making him stoop to untangle them. Already he was sweating. I let him get far enough ahead so I could catch up with him without being slowed down.

The sea bag had a rope on its neck, which enabled me to lift it up with one hand and move along sideways. I couldn't support its weight for long, so I lowered it to the floor and dragged it frequently. Ahead the bumping and scraping of the gear on the walls made a sound like an army trying to sneak into a position by night.

"Son-of-a-gun!" yelled Halmi from ahead. "I dropped my flash gun!"

"I've got it," said Skeets calmly. "I've been carrying it since we dropped down into this crack." Phil grinned at me and we both continued on.

At Last Chance we stopped for a breather. "This isn't so bad," said Phil, wiping the sweat from his face. "I had supposed we'd be belly crawling all the way."

"That's ahead of us," I said, "we'll get there soon enough."

"I brought this along but I don't see any use for it." Phil dangled a rubber respirator mask from his hand. True, he didn't need it; there was no dust to speak of. "The doctor told me to wear this," he said.

"Doc Wanger?" I asked. I couldn't imagine Doc ordering anybody to wear a contraption like that in this restricted space, especially after he had been through it himself.

"No, my doctor back in Louisville. I've just recovered from a serious lung infection, a kind of fungus growth on the lung called histoplasmosis. They almost had to remove the lung."

I shuddered at the word, *histoplasmosis*. I had seen it somewhere before, now I remembered! A New Zealand physician had studied a group of cave explorers who were suffering from a strange malady. Cave Disease was what he called it, and after further study, histoplasmosis! Phil continued:

"The Doctor didn't want me to come along, but after considerable persuasion, he said I could go, but only at my own risk. I've got some medicine along, but if you see mushrooms sprouting out my nose, you'll know what happened." Phil continued moving along the passage. We were duck-walking now. What would happen if the disease flared up? He couldn't have picked a worse place to come, for caves are notoriously damp.

We lay down on our backs in the low passage waiting for the others to move ahead. They were squeezing through into the tight part of the Crawl-way, judging by their groans. "Phil," I said, "the secret of endurance in cave exploring is to relax. Rest every muscle you can every chance you get."

He listened attentively. "Any time those up ahead stop or slow down, you stop, too, then stretch out on the floor and let yourself go."

It was time to move on. The trouble I had encountered moving the sea bag seemed small compared to the anticipated difficulty to come, for we were entering the tight part of the Crawlway. As Phil lay on the passage floor I literally crawled over the top of him with about two inches to spare. Phil tied the sea bag on my ankle so I could drag it as I crawled, since it would be impossible to push it ahead of me. I crawled a few feet, then drew the bag after me. Phil came to the rescue when it jammed against rocks and got stuck. A few moments later we came to the S Curve. Skeets, who was ahead of me now, started into it facing the wrong wall. He saw his error and backed out for another assault. When his shoes disappeared around the bend, I started after him, but too late I realized that I had neglected to relay word to Phil that we were coming to the S Curve. After we were through it I asked him how he liked this most formidable obstacle yet.

"I was so busy pushing that damn bag I didn't even notice," he said, "but it did seem to get tighter."

"The S Curve and the Keyhole are the tightest squeezes we have on the trip," I said.

Phil was complaining about being hungry, a reasonable pronouncement considering what he had eaten for breakfast. But the food we carried was to last us for lunch and a part of supper, and if we ate too much now we'd be up against it later. Words came drifting back from up ahead. Skeets relayed the message. "Hold it up on this side of the Keyhole. Halmi wants to get some pictures." We rested again, this time for about twenty minutes. I was in a position where I could talk to Skeets now.

"Whatever brought you to come down here for the expedition?" I asked him.

"When your people approached NBC about covering the event, they were ready to turn it down unless I went along. I suppose they figured it would be interesting to get the reactions of a man who has been in a cave before under somewhat different circumstances." He lay on his side looking down the length of his body at me. All I could see as I looked ahead was the glare of the flame in his reflector, and his breath rising in steam in front of it.

"Was Sand Cave like this?" I asked.

"No, just a foxhole. You had to stoop over just inside the entrance. Then it was a crawl in a wet trough, mostly downward about 45 degrees the rest of the way. But the cave wasn't as solid as this. Great blocks of sandstone were all over."

"O.K. to come ahead," yelled Jim from the other side of the Keyhole.

Skeets moved on and I followed. The rope on the sea bag cut into my leg. I hated to move again, for my ankle was sore.

Halmi's flash unit went off catching Harsham emerging from the Keyhole. "Son-of-a-gun!" he said, "this *is* a honey of a shot!" I didn't see how it could be, he had so little room. Jim Dyer was on the telephone just beyond talking to Luther Miller deep within the cave.

DYER: Luther?

MILLER: Yes.

DYER: This is Dyer at the Keyhole. We're coming in. What are you doing back there? Are you available if I need you?

MILLER: No. There are four phone lines out up over the Lost Passage, about a hundred yards from the Turnpike.

DYER: Yes, I know, but would you be available to help if I need you?

MILLER: Just a moment, I wiggled a wire.

DYER: Now wait a minute. All I want to know is are you going to stay there if I need help?

MILLER: No, I'm going on to the kitchen.

DYER: O.K. Do that then.

Jim was worried about how heavy the loads were becoming for all of us. In a little while we lay resting in the room leading to the straddle canyon. Halmi hauled out some dried fruit from his sack and passed it around. Since it wasn't part of our store to begin with, we weren't worried about eating it.

Jim was up to his old tricks. He gazed about the small room looking at passages leading off. "I wonder which way we go here?" he said with a smile.

"I know," said Halmi, "back the way we came . . . I've been warned about this!"

Jim grinned. "I'm going up through this hole. Count to ten, then come after me." With that he vaulted up through the passageway leading off from the ceiling.

". . . eight, nine, ten!" said Halmi, "I'm coming." We followed Halmi up. Jim was about a hundred feet down the passage looking back at us. "How did you get so far?" yelled Halmi.

"Like this," Jim said. He placed both hands on the canyon rim and in a moment had canyon-hopped his way back to us. Out of breath, he spoke, "How do you like that, Bob?"

"Son-of-a-gun, that'll make a great picture." We settled down on the floor to wait as Halmi rigged his electronic flash units. Jim bounded back and forth a half dozen times as the brief intermittent flashes filled the passage.

Somehow Jim managed to move the heavy amplifier along the passage. We were scheduled to make a stop at the Bottomless Pit at 3 P.M. so Skeets could make a broadcast. The sea bag rolled neatly along the floor as I worked sideways down the passage with my rump on one wall and my feet on the other. I was sweating profusely from dragging the heavy bag, a far cry from the last trip in.

"I've lost my viewfinder!" cried Halmi. No one had the time or energy to look for it. We continued onward.

Soon we rested in the muddy breakdown room at the end of the Crawl-way. Below us the water dripped. Phil offered us gum from his sack, and I lit a cigarette. I had brought four packs with me, knowing I wouldn't smoke much underground, but at the same time I didn't want to borrow cigarettes which others had brought in so carefully. When we finished the rest of the fruit, Jim suggested we wait to eat the candy bars until we arrived at Ebb

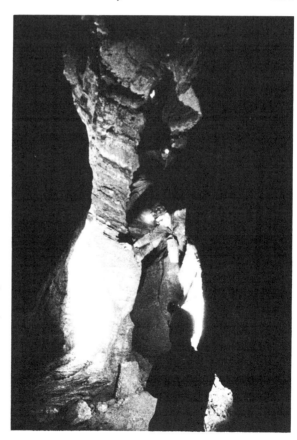

A climb down a narrow canyon leads to the Dining Room. By bracing himself between the two walls an explorer can make the descent safely. *Photo by Robert Halmi.*

and Flow Falls. For a brief moment we were quiet before starting on again.

Phil was a bit dubious of the muddy crawl along the top of the canyon as we started doubling back under the Crawlway. I moved beside him on the opposite ledge. We finally passed our gear down to the others below. The weight of the sea bag nearly staggered Jim. "I didn't know it was so heavy!" he exclaimed.

"I wouldn't trade with you, Jim," I said. The amplifier may have been

Much of the trip from Ebb and Flow Falls to Bottomless Pit requires "canyon hopping." This is accomplished by pressing hands and feet against opposite walls and moving one limb at a time while maintaining three points of support with the others. *Photo by Robert Halmi.*

lighter, but the only means of carrying it was by a handle projecting from the top. The sea bag could be bulldozed into most tight spots by grasping it any place convenient. I had long since stopped worrying about breakage, although nothing inside had shattered to my knowledge.

"Bruck, how do I get down here?" said Phil, looking with a wrinkled brow down the yawning canyon the others had just descended.

"Wait until I catch up with you," I yelled. "I'll talk you down." I was soon with him and could share some of his qualms by placing myself in the position of a newcomer to this sport. He complained that his arms were weak from all the crawling. While Jim stood below to guide him from there, Phil swung his legs over the ledge. He fished for a foothold. "Down a little," I yelled. I was in a good position to see his progress even though he couldn't. He moved his arms into a position where one hand grasped each ledge, then looked down. It was a long way, and he admitted it. Maintaining hold with one hand, he lowered his body into a crouching position and finally was sitting in a brace with his back on one wall and his feet on the other. He spread his arms again, lowering his trunk through the crevice to support all his weight on his forearms.

"Can't find a foothold!" he yelled. Jim acted before I could say anything.

"Come on down a little . . . that's it . . . stand on my shoulder." With his feet firmly on Jim's shoulders, he crouched down again, secured new hand-holds, and climbed down easily.

"I don't mind telling you I was worried there for a minute," said Phil after I had climbed down. He was still shaking.

"Next time you do that," I said, "make sure you have at least three points of support at all times."

"What do you mean 'three points' of support?"

"Well, your back might be one, your hand is one, your other hand is a third. Your feet count for two points. The whole idea is that you don't depend on any *one* point for complete support. If you do, you'll wind up in the bottom of a canyon."

"Good advice, Roger, but why in hell didn't you tell me that sooner?"

We moved on from the Dining Room along the narrow crevice covered with sharp projections. Getting the bag through was sheer torture, for it seemed to catch and hold on every projection, no matter how minute. Finally, I looked up to discover I could pull the bag along on a ledge high above my head without its snagging. Jim knocked the customary rock into the customary small hole, and the party stopped in awe as the rock plunged into deep water somewhere below.

A few moments later we stopped. "Why do they call this Ebb and Flow Falls?" asked Skeets.

Jim answered, "One night when we were in here we stopped, just as we're stopping now. The falls was just a drip then too, but all of a sudden we heard a kind of gurgling sound, like water draining out of a bathtub. Then water began to pour out in a regular cascade, setting up a loud roar —so loud we couldn't talk to each other. Then it died down to a trickle again. Since it came and went like the tides in the ocean, we named it Ebb and Flow Falls. But we've never seen it happen since."

"What do you think caused it, Jim?" said Halmi.

"I have no idea. When we came out after that trip we found that it had rained." Phil, Skeets, and I looked at the dripping water in disbelief. Could it have happened? Anything could happen in Crystal and it wouldn't surprise me. Perhaps the sudden flow had been caused by the collapse of a mud dam on a large reservoir basin somewhere above. A cloudburst might do it, whereas ordinary rain-water run-off might be impounded in the system, continuing to feed the steady drip.

The others filled their water bottles and drank deeply at Ebb and Flow. I noticed that Jim carefully removed his hat when he drank.

Skeets's broadcast from Bottomless Pit is recorded in Base Camp, then the tape will be driven to a Nashville radio station and put on the air. *Photo by Henry Douglas.*

We moved on toward the Bottomless Pit. It was now 2:00 P.M. and Skeets was scheduled to broadcast at 3:00 P.M. Our route was along the top part of a canyon, similar to others we had traveled along but wider and deeper. Jim placed one foot on one side, the other on the opposite side, and literally ran down the passage hopping from one side to the other. Skeets said he didn't dare to try that, so he crawled along one wall on the brink of the canyon. He kept pace until he came to a spot where the ledge narrowed to a scant eight inches. Here he balked and called for help.

Jim in front and I in back could see that he was in imminent danger of slipping into the canyon if he tried to move. We both scampered to him along the other wall. I braced his legs against the wall with both hands and we told him to crawl forward. Any slip would be met by Jim's hands on his shoulders and mine on his hips, flattening him immediately against the wall where he would be safe.

The only casualty during this delicate operation was Skeets' flashlight which bobbed out of his hip pocket and plunged downward ricocheting off the walls until it hit bottom. Skeets looked after it mournfully. A week in the lower levels would be no fun without a flashlight, and it could be dangerous.

"I'll see if I can get it," I said.

"Don't bother. It isn't worth it."

"You keep moving, I'll catch up." I had several personal reasons for wanting to go down there: First, I wanted to return Skeets' flashlight, and second, I wanted to have a look at the bottom of the canyon. Discoveries are made in just this manner, by people entering some unlikely place only to find a new passage. From the looks of the walls, no one had been down before; all the more reason to try it. The canyon was shaped much like the one I had gone into to retrieve my carbide lamp on the trip with Doc Wanger, so I had no qualms about its difficulty. I chimneyed down to the bottom, about twenty feet down, and found the yellow plastic flashlight lying on the floor, an eight-inch-wide strip of gravel. No passages led off that I could see, so I climbed back up and returned the light to Skeets.

We went through another small room, then up a pile of breakdown. At the top, a tangle of wires indicated we were on the supply route. A light appeared down the passage moving toward us. It was Bob Handley carrying a field telephone. Electrical pliers protruded from his pocket, and he carried a small coil of telephone wire in the other hand. He stood about six feet tall. A tattered vest flopped loosely around his chest, revealing a faded gray flannel shirt beneath. He wore gray cotton pants and a pair of miners' shoes with steel toes now polished to a high luster from being dragged through Crystal Cave. His appearance belied the fact that he is acknowledged to be one of the best cave explorers in the country. A draftsman from West Virginia, he is credited with having explored most of the Organ-Hedricks Cave system, a vast cavern network under West Virginia.

"I'm going to hitch up the phone at the pit, Jim," he said in a mountain drawl. With that he turned around and trotted off into the blackness.

We were in a low passage now, about five feet high and eight feet wide. Rocks covered the floor but travel was comparatively easy. Jim stopped at one point to indicate the place where Floyd Collins had scratched his initials on the wall.

Skeets asked, "Why did Floyd do that?"

"Floyd was very proud of his discovery," explained Jim, "and he marked his initials so others could tell he had been over this route. It's a funny thing, though. Floyd got this far, but we know he never got as far as the Bottomless Pit , only about three hundred feet farther on. Why he didn't go on is beyond us."

"Well, how do you know he *didn't* go on?"

"We discovered the Bottomless Pit one night by coming this same way. When we got there, we found none of the usual signs that someone had been to the place before, none of the dust was disturbed and none of the rocks had marks on them as though anyone had traveled over them."

"But if he discovered the Lost Passage originally, how did he get there if he didn't come this way?"

"Ah, now you're catching on! Back by Ebb and Flow Falls there are a number of passages leading off. You didn't see them because you were too busy following me. Floyd's route to the Lost Passage lay through a long and rough crawlway called the Flint Crawl. It connects with a pit called Floyd's Jump-Off—you'll see it later in the week—which leads to Mud Avenue and from there into the Lost Passage." The three reporters puzzled over this confusing list of names, but Jim indicated that we'd better move on.

At two thirty we reached the Bottomless Pit. I was weary from carrying the heavy sea bag, so I placed it on the passage floor and stretched out, placing my head on it. Jim called for Skeets and Phil to come up with him. I could see their backs as they gazed out on that void called the Bottomless Pit. Jim picked up a rock and hurled it into the hole. Skeets counted to four and it clattered on the bottom.

"You mean we have to cross that?" asked Skeets.

"Isn't there some *other* way around?" added Phil.

"Son-of-a-gun! What a picture!" said Halmi.

Bob Handley called to us, and in a moment had crossed the Pit. "Been putting wires up off the trail," he said. "Let's hook in and see if we can rouse 'em upstairs." He went to work stripping wires with his knife, splicing one lead from a telephone terminal, then the other. He cranked the ringing magneto furiously, then listened. He cranked again—still no answer. "Probably got the wrong leads," he said. He unwound the splices and tied into the third wire. This time his cranking evoked an answer from those in Base Camp. He said we were at the Bottomless Pit and that he'd been working on phone wires all day. He handed the phone to Jim who made a short report on our progress.

Skeets took the phone next. "Hello, Frank! . . . Yes, we're all right . . .

what? . . . no, we have the amplifier right with us . . . no damage that I can see. It seems to have survived the trip fairly well . . . me? I'm tired, very tired, and we've had to go easy on the food because of the supply situation . . . all right, Frank . . . we'll sign off now and hook in the amplifier . . . Frank . . . stand by on this line. We'll try to contact you in five minutes . . . O.K. . . . goodbye, Frank."

Skeets handed the phone to Jim, who placed it in the case. Bob Handley was already stripping wires for splicing in the amplifier. Skeets opened the case and took out some strips of foam-rubber packing. Next he removed a small microphone and a set of earphones. He slipped these on his head, and turned the amplifier on. A red light glowed warmly. Two minutes to contact. "I can hear them up there," he said. "Getting lots of cross-talk on the line, though." Handley listened and nodded.

"Suggest that the switchboard cut everyone off when it comes time to broadcast," said Handley in a low voice. Cross-talk occurs on telephone lines when several circuits are running close together. Conversation carried on one line actually leaks over to the other. Since our phone system was primarily for communication and sound quality was secondary, it didn't make too much difference to us, but it would lower the sound quality by giving it that authentic far-away sound usually associated with documentary broadcasts.

"Hello, Frank . . . I hear you . . . are you picking me up . . . what's that? . . . louder . . . I can hardly hear you either . . . yes, I've moved up the gain . . . no, it doesn't do any good . . . shall we try it anyway? O.K. Give me a time check . . . two forty-eight exactly . . . O.K."

Skeets checked his watch. The second hand rolled around into the 12 position, and he started speaking. He talked for about a minute describing the trip in, then was interrupted. He listened intently, then took off the earphones.

"We're going to abandon the amplifier," he said. "We seem to be able to get more power and a clearer signal through the phone itself." Bob Handley stripped wire for the third time and in a few minutes had the phone working again.

During this time, Bob Halmi had unsheathed his cameras and was taking pictures of the proceedings. Skeets lay prone in the passage and resumed contact with the staff of radio station WSM in Nashville. "I'll give it to you again. Ready?" He began speaking drily, the voice of a tired man, weary with exhaustion. He spoke of the ruggedness of the trip in, how he'd dropped his flashlight and had given it up for lost; then he verbally patted the backs of the guides. That made Jim and me feel good. After completing the broadcast he spoke to his wife, who seemed concerned for his safety. He assured her he was in good hands, then he hung up.

I stood up and began to move about since the period of inactivity had left me chilled. Phil got up from where he lay and moved to the phone. Someone at Base Camp wanted his impressions of the trip in. He looked tired as he spoke, and his voice confirmed it. He received a merciless grilling by somebody on the other end who demanded to be told about every inch of the passage. We grinned when he answered a question: ". . . I admit it, I am pooped!" We felt the same way, and were pleased to learn later that the *Courier-Journal* chose these very few words for a large headline on page one. They summed up a feeling that all of us would voice in the days ahead. There was a moment of silence, and then a noise.

A clatter of rock in the passage behind us indicated someone was coming toward us at a rapid rate. Lights appeared, three of them. Bill Austin came into range and, behind him, two CBS television newsreel cameramen. All of them carried bulky boxes, and one clutched a portable floodlight for making movies. They paused, momentarily out of breath.

"We're going down to the other side of the Turnpike," said Austin, "and get set up. When you come through, the boys want to take some pictures." "The boys" puffed and panted from the murderous pace at which Austin had led them through the cave. But there was not time for them to bemoan their fate. They were across the pit and out of sight before we packed our gear and began to move. Dyer and Halmi leaped over the front edge of the pit to a narrow ledge of crumbling limestone. They squeezed through the crack between the big poised boulder and the wall. Viewing this, Skeets showed his dread at making the leap, so to be on the safe side I insisted that he allow me to put a safety line on him. I secured a bowline around his chest and crossed to the ledge while Phil stayed behind to help if needed. I wrapped the rope around my body and braced my feet against the boulder and the wall. If he should slip now, he would travel only a few feet at the most before the rope would break his fall. He didn't depend on the rope, however, but crawled carefully over to me with no hesitation, like a man crawling on eggs. I launched him into the crack behind the boulder, then crossed back to allow Phil Harsham to continue.

When I arrived at the main part of the pit, Jim was already "talking" Phil across. "Keep your hands on the wall . . . that's it . . . doing fine now . . . don't hang on the phone wire! . . . you've almost got it . . . stretch out your right hand now . . . I've got you, you're O.K. now." Just then Phil looked down into the pit he had just crossed; it occurred to me that he had obviously avoided looking down while crossing. He shined his flashlight down into the pit. "I'm glad I didn't see before what I'm seeing now!" said Phil. The walls of the shaft went straight down into blackness. I hooked all the equipment on the cable and started over, sliding the equipment as I went. Jim removed it on the other side.

Nearing the Turnpike, Phil tossed one of his rucksacks ahead of him as we had done so often in the Crawlway. It rolled, then disappeared in an opening. "I'll get it," he yelled. Before I could warn him he dived into the hole. He returned with the sack. "I was lucky, I guess. The sack was resting on the brink of a deep pit through that hole. If I were less tired and had been traveling faster, I'd be at the bottom." Phil was perspiring.

Jim Dyer looked back sternly. "I think that teaches you one lesson about Crystal Cave. You don't throw your stuff where you can't see it, and you don't charge into a hole unless you can see where you are going. There are just too many passages and crawlways that open into deep pits."

A blaze of light ahead marked the Turnpike where the newsreel men were set up ready to go. Jim crawled through and Halmi after him. Skeets took one look at the ledge, flanked by a solid wall at one side and a deep pit at the other. He wouldn't go headfirst. Instead, he carefully turned around and backed in feet first, with Jim shouting directions. All this was duly recorded on the motion-picture film.

The cameramen and Austin turned back at this point and we headed on. The size of the Turnpike Room amazed the reporters, for we had entered nothing like it since we left the commercial route. They commented on the slabs of breakdown covering the floor, huge rocks, some ten feet by ten feet by two feet thick. They wondered if there was any likelihood of more falling down. Jim said he didn't think so, since none had ever fallen down of its own accord while anyone was in the cave.

Our route took us to the end of this ballroom-like cavity, through a crawlway, around the brink of another pit, and finally down into a room with a red sand floor. The phone wire terminated here. Bob Handley was crouching on the ground splicing in the telephone. A light appeared from a hole in the floor and Roger McClure stood beside us. "Welcome," he said, "to our humble quarters."

"Is this Camp One?" asked Skeets.

"No, it's down the line some distance yet. This is as far as the phone goes though, so you might say Camp One extends from here to the Lost Passage," said Roger.

Bob Handley looked up. "We've got three lines to this point and I'm checking them out now. We're getting more wire down here soon. The wire party refused to believe that all those thousands of feet of wire they dragged in wouldn't reach to the Lost Passage, but I've called out for some more to be brought in. With luck, we'll get the lines all the way down sometime tonight." Handley resumed testing the lines.

"I come up here every hour to report in," said McClure.

"Roger is in charge of setting up Camp One," explained Jim. Bob Handley made contact on all three lines, then started back to meet the incoming wire party.

"Won't he get lost?" asked Phil.

"No," said Jim. "He's a pretty good man when it comes to finding his way around. He's been running back and forth over this route all day."

Rog McClure took the sea bag from me and I took several of the rucksacks. We traveled across a low-ceilinged room and then ducked into a hole in the opposite wall. Jim pointed out the crack in the floor that led to the Lost Passage. We paused for a quick picture of Jim climbing into the crack, then went on dropping down into a spring basin beside the trail. A low muddy passage brought us into the Lost Passage at last.

Skeets, Halmi, and Harsham were unprepared for what met them. They had been used to a certain level of illumination from combined headlamps, but now the level dropped sharply. The close walls which had previously reflected the light gave way to a very wide and high passage. The headlamps hardly pierced the gloom. They stood agape at this passage running into blackness on either side of them.

"As big as a railroad tunnel," said Phil.

"Unbelievable," whispered Skeets.

"Son-of-a-gun! SON-OF-A-GUN! I've got to get a picture of this. Where's that bag?" Halmi shouted.

"Isn't anybody hungry?" said Jim. "You've got all week to get pictures of this. It's going to be your new home."

Nobody volunteered to linger; we were all very hungry and very tired. We moved along the Lost Passage in silence, watching the play of lights against the wall. At a bend in the passage we saw candles burning on tops of old tin cans. At one side of the passage was a trench with all the cooking gear spread out on one edge of it. Back of the trench was a flat rock propped up against the wall. On it, in neat carbide letters, was the inscription, "McClure's Slop Shop." We all laughed heartily, for it was as though we had just climbed to the top of Everest and found a stone marked "Hillary's Hamburger Heaven."

Our laughter was short-lived, however, for the food situation was serious. Our meager stockpile consisted of four sacks of peanuts, four candy bars, and half a bag of dried fruit. The advance party at Camp One had already eaten most of their rations, so the only other food that remained was a half pot of coffee, a small pan of oatmeal, and some rations I had brought along.

We nibbled on these tidbits, finishing them with an unpleasant gnawing feeling in our stomachs. All the sleeping bags were occupied by members of the advance party. Jim woke up three of them who had been searching for a site for Camp Two all day. He hated to do it, but Skeets Miller, Phil Harsham, and Bob Halmi all looked as though they might collapse at any minute. It was 7:30 P.M. when the three staggered down the Lost Passage to their sleeping bags.

Jim Dyer, Roger McClure and I sat talking for the next half hour. McClure told us that members of the advance party had traveled out B Trail to look for a campsite for a second outpost, and had come back discouraged, saying that they had found no suitable place that was near water and had room enough.

We were silent for a few minutes, too tired to carry on a conversation.

Roger Brucker had had K.P. the day before he entered the cave; when he arrived in Camp One, Jim Dyer immediately put him back on K.P.
Photo by Robert Halmi.

A light appeared down the passage and moved toward us; it was Bob Halmi. "I can't sleep," he said. "Thought I'd come back and sit with you."

Roger McClure looked at Jim. "In that case, I think I'll take that sack. I haven't had any sleep since we came in yesterday morning." McClure put on his battered white hard hat and stumbled off down the passage. I wished I could have followed.

"Brucker," said Jim, "I'm putting you in charge of the kitchen. Keep things under control and when anyone comes, try to scrounge up something to feed them. Supplies ought to be in here soon." Jim went over to one wall and sat down against it. He buried his head in his knees to try to sleep.

Halmi got up and began to remove his coveralls. "Where can I hang these up?" he asked.

"Hang them up? What for?"

"I want to dry them out. They're damp, too clammy for comfort."

"Hang them on the ledge over there. You can put a rock on the neck so they won't slip off." I watched as he walked over to the wall in his woolen underwear, placed the coveralls on the ledge, picked up a rock from the sandy floor and put it in position. Then he stumbled back to the feeble warmth of the candle. I knew he'd be in for a shock when he put them back on. They'd be just as damp as when he removed them.

"There doesn't appear to be much gasoline here," I said, "but let's light the lantern anyway. Gasoline is supposed to come in with the first support team." I poured a little into the tank of the lantern.

The white glow of the hissing lantern gave a new perspective to the Lost Passage. Our range of vision had been limited to dim vistas by carbide lamps, but now we could see what Camp One really looked like. The passage was about thirty feet wide at this point and continued that way as far as we could see in either direction. The ceiling over our heads was about seven feet high, sloping down to about six feet at the walls. It was light gray, crossed here and there by joint lines. The floor of dark brown sand undulated gently like swells on a calm ocean. It almost looked man-made because of its regularity and uniformity. Jagged rocks, so prominent in other parts of the cave, were absent here. All this made for a kind of order and peace in an otherwise chaotic cave, and the warm friendly light of the candles and lantern gave it the real touch, for without them the Lost Passage would be just so much blackness. Sleeping bags were strewn here and there, wherever their tired occupants had found a smooth spot in the sand.

The personnel build-up at Camp One was well under way. From Base Camp we learned that parties bringing in supplies were spread out through the cave, all crawling toward us. Some would stay the rest of the week, and some would return to the surface immediately to bring in more supplies in Gurnee cans and sea bags. This was the start of a round-the-clock operation calculated to stage a great assault on a baffling cave.

After our departure from Scotchman's Trap the Base Camp Log clearly filled in the picture of the migration:

Departures: 1252 hours. Barr, Richter, Gehring, Shoptaugh—Support to Camp One.

1427 hours. Thierry (leader), Blakesley, Rogers, U. E. Lutz, Welsh. 2 sleeping bags, 3 meals for 5, eating utensils, personal gear for four.

1815 hours. Lehrberger, Smith, Wilt, Bill Wanger. Food and sleeping bags to Camp One.

2000 hours. Perry (leader), Dymond, Klein, Sawtelle, Spence, Courneyer. Equipment and supplies.

By Monday afternoon the complement of Camp One would be complete.

FOURTEEN-HOUR NIGHTMARE

FOUR LIGHTS AND VOICES UP THE PASSAGE, the way we had come in, sig-naled the arrival of the first support team. I looked at my watch; it was 8:00 P.M. Bob Richter and Tom Barr were leading; behind came John Gehring and Glenn Shoptaugh, both interns, carrying or dragging supplies.

"Welcome to Camp One," I said.

"Not much of a camp," Bob replied. "I expected to see a blaze of lights and a flurry of activity. Where is everybody?"

"Sacked out, pooped. They've been on the go all yesterday, all last night and today."

"Well, we've brought you some scientific stuff," said Richter. He walked over to the wall behind the kitchen and dumped out a sea bag full of aluminum soil-sample bottles. Soil-sample bottles! And the need we had for food!

"Just where *is* the food?" I asked.

"We only brought one Gurnee can load of food," said Barr. "We heard you had plenty of food. Say, we could use some. We haven't eaten for six hours." He eyed the pot of oatmeal bubbling on the stove.

"I thought you were supposed to bring your own food for the trip in and the trip out. That was my understanding," I said.

"Yes, well, we got hungry and ate it on the way in," admitted Barr.

I ladled out a cup of precious raisin oatmeal, then went over to unpack the food. It was jammed inside a torpedo, now battered and dented from the trip in. One of the seams in the galvanized metal had begun to open. They won't hold up for many more trips, I thought, but the food inside was

118

in good condition. I stacked it on the floor between the wall and the trench, trying to arrange it in neat piles.

"Tom and I are going off for a little while to do some scientific work," said Richter. They gathered up a handful of tiny bottles and wandered down the passage. I collected the cups and dirty pots in one place, then went to the spring to get water to boil for dishwashing. In a half hour I had Camp One cleaned up ready for business. Halmi watched all this in silence, occasionally asking what I was doing or where I was going.

At nine o'clock we heard more noise. Another supply team had entered the Lost Passage, I thought. I put oatmeal and coffee on to cook and settled back to wait for the visitors. We were surprised to see no supply team after a while, so I walked up the passage to see who was doing the talking. Bob Handley was still at it! He and Hugh Stout, a graduate student and assistant geologist for the expedition, were stringing wire down the Lost Passage. They held a reel between them and walked for about a hundred feet, then went back to pick up a second reel. They repeated this process after each hundred feet or so. They had run out of the third wire earlier, and were now working furiously to bring at least one circuit to Camp One. I helped them for a while, then when their wire ran out a hundred feet short of Camp One, I invited them to have something to eat.

They gulped down the oatmeal and coffee, finishing up with dried fruit from our augmented supply. They said very little because of fatigue, but announced that at last report several more supply teams were on the way in. Bob and Hugh left to find the phone and see about getting more wire to finish the circuit and to bring in an additional circuit.

At ten o'clock Jack Lehrberger arrived leading a support party loaded down with food like a caravan. From that moment on our food worries were eased. Halmi and Dyer got up from where they had been sitting to greet the newly arrived team. Sweat rolled off the team members' faces, dropping on their clothes covered with dirt. It looked as though they had made a fast trip, because perspiration had soaked through their clothing and caused cave dust to stick. They were soon eating hot food from the stove, talking and laughing jovially about the trip in, and now and then one would gaze at his surroundings with his mouth open as if in disbelief.

Luther Miller was awake now, wondering what had happened since he had gone to bed late in the afternoon. He was amazed by the mountain of food comprising Camp One's store. I briefed him on the details of the communications situation, telling him that Handley and Stout were back along the trail somewhere straightening out the wires. He gulped down some oatmeal, then left to find them.

Around eleven Sunday night a large party arrived, most of whom were to stay at Camp One for the remainder of the week. Earl Thierry, the chief surveyor, was there. Audrey Blakesley and Lou Lutz of the climbing team looked ready for action. Hundred-foot lengths of new nylon rope were coiled around Lou's shoulders. A karabiner loaded with pitons jingled at his belt. Nancy Rogers, whose specialty is bat-banding, looked fresh and eager. "Didn't see any bats on the way in," she said. George Welsh, guide for the party, grinned at the excitement, although he was scheduled to go out in a few minutes. I looked around for Bob Halmi to see his reaction to this activity, but he was nowhere in sight. I found out from Jim that he had taken Luther's sleeping bag and was already snoring loudly from his berth near the spring.

To ease the confusion, Jack Lehrberger gathered his supply team together and started out with George Welsh and Tom Barr. The party trooped off down the passage, passing Bob Handley coming toward us laying wire. They exchanged greetings loudly.

Bob hooked more wires on the end of the line, then brought the telephone into camp. It was midnight when he cranked the phone to contact Base Camp from deep in the bowels of the earth. If calling from the Bottomless Pit was a thrill to me, the magic words "Camp One calling" were nothing short of a miracle. Bob talked for a while and then handed the phone to Jim Dyer.

Albertine Talis was on the other end. She wanted Jim to read the log so it could be tape-recorded. They chatted for a moment, then hung up, there being nothing urgent to communicate.

When Jim left Nancy Rogers and Bob Richter in charge of the phone, I went to talk with Roger McClure. He, Earl Thierry, Audrey Blakesley, and Luther Miller were planning to start immediately on the survey of B Trail, one of the main passages off the Lost Passage. Dyer told them how to recognize the point at which Bogardus Waterfall Trail went to the left, and the party departed.

About 2:30 A.M. on Monday somebody at Base Camp wanted a complete inventory of the food stock in Camp One. Richter busily listed each item and phoned it in. Nancy Rogers was on the telephone now. By the tone of her answers, I knew something was up. The few of us remaining in the area stopped washing dishes to watch her as she finally hung up and entered the result of the call in the log. Then she explained the situation: A party had started in under the leadership of Ken Perry. He had with him a telephone destined for Camp One, and had spliced it into the line to call Base Camp with the message that his party was making slow progress because they were so heavily loaded. That was our first indication that the fourteen-hour nightmare was taking place a half-mile back on the rocky route connecting Camp One with the outside world.

The 14-hour nightmare starts when the heavily laden party stops in the entrance passage to take inventory. Left to right: John Spence, Ken Perry (leader), Lou Dymond, Ida Sawtelle, Don Cournoyer, and Marguerite Klein. *Photo by Marguerite Klein.*

It had started at precisely 8 P.M. Ken Perry assembled his party in front of the supply tent and checked their inventory against their loads:

Ken Perry, Party leader—Lineman's tools.

Lou Dymond—One EE-8 telephone.

Marguerite Klein—3 telephone transformers, 2 coils wire.

Ida Sawtelle—1 Gurnee can of meteorological supplies.

John Spence—2 bags of supplies, camera box.

Don Cournoyer—Food for two meals.

Each person was loaded to full capacity but uncomplainingly they marched down the hill into the cave. At the phone in Grand Canyon they rested, taking the opportunity to report to Base. Perry learned to his astonishment that orders had been issued for his party to pick up two additional sleeping bags along the commercial trail. He bitterly complained that his party couldn't possibly carry the additional load, but agreed to try. He was right; the two rolls were the last straw, and at the Valley of Decision, Perry tapped into the phone line again, turning the crank. Once more he requested permission to leave the sleeping bags. After hot words with the person on the other end, he abandoned the bags at the side of the path.

At Scotchman's Trap the party stopped again, this time for a half hour, while Perry again tapped into the line. As a side mission on this support trip he had been given the job of making frequent checks to see if he could discover which pair of phone wires was which, because there had been some difficulty in bringing a second phone into Camp One. Once a second phone was installed, a third phone set could be patched in on a phantom circuit.

The party labored heavily to swing their loads into the Trap, making several trips from the commercial route to the winding canyon before all the load had been relayed. Now they moved forward, more slowly than any party had before. Fatigue was beginning to take its toll at this early stage of the trip.

After what seemed an eon, they rested for a half hour at Last Chance, barely able to prop themselves up on the gypsum-lined walls. Sawtelle suffered, not from fatigue, but from the slow progress of the party now only about four hundred feet into uncharted cave. They became chilled waiting for second wind. They rallied their spirits and moved again.

Progress was painfully slow, partly due to the burden they carried and partly due to comparative inexperience on the part of two members. Inch by inch they struggled through the tightest part of the Crawlway near the

At Scotchman's Trap Spence checks his two cameras and two flash guns before going ahead. *Photo by John Spence.*

S Curve. Again they rested and panted. Ida had been moving ahead of them, picking up loads, moving them forward, then returning for more. She grew concerned over the party's condition. Dymond felt ill; his face was ashen and he complained of a headache. Perry and Spence had gotten only a

John Spence hauls a Gurnee can loaded with meteorological equipment through the Crawlway. *Photo by C. N. Bruce.*

few hours' sleep the night before, resulting in eyes that were red-rimmed. Marguerite had long since ceased talking and looked blankly into space. The Gurnee can full of meteorological supplies lay abandoned several hundred feet toward the Trap.

Spence guarded his heavy box of camera supplies as though it mattered more than anything else in the world. He valiantly unpacked his camera during each of the increasingly lengthy rest stops to take "one more masterpiece" for the record. Perry realized their predicament and spliced the phone into the line. He reached the switchboard, demanding to be put through to Joe Lawrence. The switchboard operator explained that Joe was to be disturbed only in case of an emergency. Perry felt it certainly was one, and the call was put through to the Mammoth Cave Hotel.

When Ken Perry feels his party cannot reach Camp One with the load they are carrying, he splices a phone into the line and calls Lawrence to report the condition of his party. *Photo by John Spence.*

A bellboy knocked at Joe Lawrence's door. "You have an emergency call from the cave. You can get it at the desk." Joe's heart pounded in terror as he pulled on his trousers. Emergency! He had dreaded that word. Had someone been injured?

Perry outlined his party's plight, requesting permission to abandon supplies. Lawrence told him that as a party leader he should do whatever was necessary for the safety of his party. When Perry asked whether or not to press on toward Camp One, Lawrence again left the decision to him, since a leader would, under most circumstances, be best able to assess the strength of his own party.

Perry polled the group; the consensus was to press on. In six hours they had traveled nine hundred feet. Reason became fuzzy as time wore on. They moved a few more feet, then halted. A support party was coming back. How far to Camp One? "Why, you're only about a fifth of the way there," said one member. "They don't have any food down there anyway," said another. A weary group brewed coffee in the small room at the entrance to Straddle Canyon, attempting to put new life into those who had already reached their endurance limit. Dymond turned his back and vomited. A call to Group Leader Dyer at Camp One gave them new hope: he would send out a relief party as soon as rested personnel were available. Ida started the group singing a song, a feeble effort that died after the first few half-hearted notes.

In another hour they raised themselves to their elbows and continued to follow the phone wires. At this early stage the supply route was not well marked. They had been told to follow wires. After all, what could be more logical? Three passages led off from a room; a tiny narrow crack to the right, an improbable hole up through the ceiling, and the third, a low left-hand passage with phone wires leading into it. They moved into the bottom of Straddle Canyon, a passage pinching down to barely eight inches wide. They didn't know that the communications men had laid the wires in the bottom to get them out of the way. But they moved forward painfully, until the leader decided they had missed the route. They chimneyed up to the top where the passage had been well marked by scores of footprints.

In another hour they reached the end of the Crawlway. Telephone wires led through a tiny crack, Jack Lehrberger's short cut. Somehow they squeezed through and arrived at Ebb and Flow Falls where they drank

◄ Perry's support group brews coffee to raise sinking spirits five hours in on the 14-hour nightmare. Left to right: John Spence and Don Cournoyer. *Photo by Marguerite Klein.*

deep, then stretched out. Spence and Klein now felt they could not go on.

Perry spliced into the phone line again, learning that Jim Dyer had dispatched help ten minutes ago. Perry went ahead, meeting the relief party at the Bottomless Pit just as they had completed rigging ropes for a safe traverse. He returned with the news and the exhausted party reluctantly agreed to follow to the Pit since he reassured them that Camp One was not far beyond. At this point Perry put forth the plan that he should take Lou Dymond out the way they had come because of his poor physical condition. Dymond droned in a monotone that he had come this far under his own steam, and he would certainly be able to make it to the Lost Passage.

Charlton, Loyd, and Gurnee, the relief party, debated which route would bring the group to Camp in the shortest possible time. Charlton won out, leading them in by way of the bottom of the Bottomless Pit, a tortuous crawlway ending in a fifteen-foot chimney. Too tired to care, they stumbled on, following their two guides to the warmth of Camp One.

They straggled in. Don Cournoyer was first, looking like a zombie, with deep-set eyes and a drawn face. His sweat-stained coveralls were torn in several places. A white area on his back marked where salt crystals had formed. He sat down and stared straight ahead. The rest came, John Spence, lugging his camera equipment and babbling about a wonderful shot he had made while the party had stopped for coffee. From the looks of them, they needed more than coffee.

Marguerite Klein sat down on the edge of the trench. For five minutes she was motionless, looking straight ahead almost without blinking. Then she began to tremble, from the cold. I took off my sweatshirt and handed it to her. She refused it, saying she was all right and would feel better after eating. Firmly I told her to put it on; this time she did so without argument, while I thrust a steaming cup of coffee into her hands. Lou Dymond wandered about the camp aimlessly for a while, then lowered himself to the cold sand floor. We sent him back to a sleeping bag immediately. Ken Perry came in next, then Ida Sawtelle, Russ Gurnee, Ackie Loyd, and Roy Charlton.

Perry bravely tried to stand while he ate, insisting all the time that we take care of the others. I sent someone up the passage to clear out sleeping bags for the group. When I turned around to find Ken, he was lying on the floor.

"I know there aren't enough sleeping bags," he said. "I'll just sleep here." He said it in a monotonous tone of voice, halting, jerky with fatigue. His eyes were half closed. In a few minutes we had everyone except Ida assigned to a sleeping bag.

Ida was the antithesis of the others. She was wide awake and immediately volunteered to help with the kitchen chores. I had been caving with her before and I was glad she was with us.

Doc Wanger was on the phone from Base Camp wanting to know in what condition the party had arrived. I told him Lou Dymond had complained of nausea and headache, but was now sleeping. Marguerite was suffering from uncontrollable shivering, even though she was in a sleeping bag. John Spence had bruised his leg, and the party was suffering from general fatigue. Doc said to make sure all of them ate something and got plenty of sleep and were not disturbed. He was planning to come in with a party at 5 in the afternoon.

The fourteen-hour nightmare was over, but its after effects would be with us for days to come. Marguerite Klein lay wheezing with asthma, in a sleeping bag, where she would remain constantly for the next two days. Three other members of the party had been on the verge of collapse, dangerously close to peril. What could account for their unbelievably slow trip in? First, the physical condition of some had not been up to par prior to their departure, although all had been able to pass the physical exam. It was generally agreed that the party had been overloaded, and saddling the party with the additional assignment of checking the phone lines hadn't helped the situation. The lack of an experienced Crystal Cave guide had led them into difficult and untraveled routes. Morale was low as a result of arguments with Base Camp personnel over Perry's decision to abandon the load. Further, the news that they had penetrated only one-fifth of the passage to Camp One, accurate almost to the foot, broke down the inner will to go on, though pride and a sense of obligation would not let them turn back.

Profiting by this hard-learned lesson, the leaders vowed that never again would a party average only seven hundred feet an hour. From now on, all loads going into the cave would be weighed on a scale. Parties would be limited to a maximum of four members. Support teams would go in only when rested.

Supply teams began to thunder in on schedule dragging battered Gurnee cans. On seeing the condition of the metal containers that were our lifeline, Russ Gurnee realized that replacements were needed, and fast. Recruiting Al Mueller, he headed back toward the entrance with haste. At Base Camp he found willing assistants to accompany him into town for the fabrication job.

The local tinsmith shook his head. No one could make in an afternoon such objects as their plans called for. Gurnee solved the problem by renting the man's entire shop, shooing the workmen out, and laying out the patterns himself. Mueller and Gurnee, working swiftly, were even able to improve on the original design by forming a straight tube instead of a tapered one. They reinforced the nose rings at the end of the cone, one of the earlier can's weak points.

While Russ cut and shaped the metal, others sought canvas for covers. An apparently fruitless search for canvas ended in a shoe repair shop, where they were able to buy what they needed, then use the shop equipment to put in the drawstring grommets.

As supply teams became more familiar with the cave, the five-hour one-way trip was gradually reduced to three hours, then to the astonishment of all, one hard-driving party thundered into Camp One in one hour and fifty minutes!

Joe Lawrence's prediction had come true. The ultimate success of the smooth-running support operation, once the difficulties had been ironed out, was due in large measure to the continuous efforts of some of the best cavers in the group. In recognition of their efforts, he would allow them to exchange places with those in exploring parties as the week progressed.

Supply Officer Alden Snell was always busy. If he wasn't lending a hand in the Base Camp kitchen he was making the daily purchases for the expedition or filling a request for the explorers down below. *Photo by Glen Shoptaugh.*

11

TOP OF FOOL'S DOME

D ENIED THE PRIVILEGE of getting sleep, at least for some hours to come, I moved to the phone where Hugh Stout stood. We talked about the sleeping bag situation for a few moments until he told me he knew where an extra sleeping bag was cached along the commercial route. I turned abruptly from him and cranked the phone vigorously. Then I entered the time in the log: 11:35 A.M. Monday, February 15th. Base Camp answered quickly, connecting me with the supply officer.

"Snell? Brucker in Camp One. . . ."

"What can I do for you?"

"There's a sleeping bag hidden away behind Floyd's wheelbarrow on the Helictite Route. There's an air mattress inside, plastic."

"Yeah."

"Can you have the next support team bring it in when they come?"

"Will do, Bruck. Anything else?"

"Just a minute . . . Stout says there ought to be another mattress kicking around the supply tent. Better include that if you find it."

"O.K. Bruck, anything else?"

"No, thanks, Alden . . . supplies are coming along in great shape."

"O.K. I'll sign off now."

Often, using the telephone was like pulling a finger out of a dike. The answerer would get stuck there answering a flood of calls for ten or more minutes. This was one of those times. No sooner had I hung up than the phone rang again. Joe Lawrence wanted to know the condition of Ken Perry's party so I filled him in, telling him that all of the group with the exception of Ida were now asleep. I looked over at Ida brewing coffee;

129

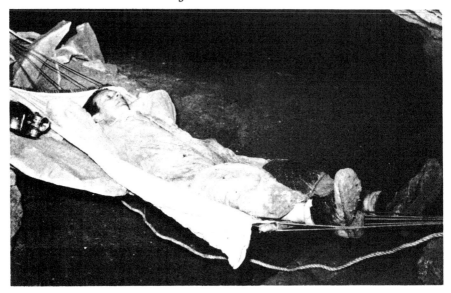

George Welsh got a few moments' rest in his hammock before he
tumbled to the bottom of the trench. *Photo by Glen Shoptaugh.*

she looked chipper and still fresh after the ordeal she had come through.
The press representatives had been taken in tow by Dyer about an hour
ago, so he could show them the gypsum flowers. No doubt they would
come back raving, as everybody does who sees the flowers for the first time.

Next, Doc Wanger called to find out how Ken Perry's party was faring.
I told him I had followed his instruction to the letter. He wondered if there
was anything special we needed since he was starting in with a support
team at five in the afternoon. We needed nothing out of the ordinary, so I
told him to bring whatever load Snell gave him.

At noon the phone rang again, someone wanting Dyer to call the switchboard on his return from the gypsum flowers. Then I decided that if I
moved away from the phone, others would have to answer it. The telephone
was quicker; it rang again. This time I wrote a note in the log; cryptic
sentences that meant little to me but would have meaning to the wire party.

12:17   Douglas leading in with three rolls wire. Three in party.
        Stout and Handley are to meet party at end of Crawlway.
        Note to Ken Perry: Call Bob Lutz on arising (RWB)

This time I made good my escape. The most inconspicuous refuge proved
to be in plain sight of all, in the kitchen washing dishes. Not many people
were in camp that time of day, but soon they would be drifting back de-

manding to be fed and complaining about the food but not the cooking. Nancy Rogers had arranged our storehouse in neat order. There were about a dozen packages of dried fruit, some cheese bricks, assorted cans of coffee, tomato juice, roast beef, boxes of cereal, sugar, and even a bunch of carrots. It covered an area about as big as a dining-room table between the edge of the trench and the wall.

Audrey Blakesley came over to help, seating herself on the edge of the trench. There was a loud "crack" and Audrey tumbled to the floor of the trench in a tangled heap. "What are you trying to do, bust up our kitchen?" yelled Nancy. Audrey picked herself up and surveyed the remnants of the ledge that had broken under her weight. Later in the week others sat on the ledge and found themselves in the bottom of the trench, in just the same way. George Welsh, however, walked off with the funniest trench laugh of the week, when he set up his rope hammock for the benefit of Bob Halmi's cameras. In the Lost Passage, the trench offered the only convenient place to suspend it, so he carefully tied both ends to boulders on either side. He

At the south end of the Lost Passage gypsum flowers grow from the ceiling in one of the finest displays yet discovered. Explorers remove hats to prevent breaking any of the fragile formations. *Photo by Glen Shoptaugh.*

climbed into it slowly to avoid tipping over. Electronic flashes blinded by-standers for a moment. Then George stole the show by plummeting down to the soft sand in the trench bottom. One of the end ropes had broken.

The press party was back from the gypsum flowers acting like a chamber of commerce for the cave. Harsham retreated to a quiet corner to compose adjectives befitting the wonders they had just seen. Skeets Miller went back down the passage to record his impressions.

> "In an earlier broadcast on WSM Mr. Dyer heard that I was dis-appointed not to see the ceiling that was out of sight. After we got to the area he pointed out that it was not so much the height of the ceiling that bothered us, but that it went for great distances horizontally. To convince us, he took Bob Halmi, Phil Harsham and me into one of the most beautiful sights any of us had seen. This was a long walk up and over the fallen ceiling. It had fallen many years ago. After we had passed this Mr. Dyer had us remove our hats and told us to keep our heads low. It was almost like en-tering a sanctuary. Here, on the ceiling, was an orchid paradise. These were delicate gypsum formations in the form of orchids de-fying the laws of gravity. They hung from the roof in clusters. They were of indescribable beauty from a pure white to a reddish brown. If this cave is ever opened up for the tourists, the floor will have to be lowered because it is impossible to expect them to re-build themselves in any lifetime should any of them be broken. It has been estimated that these have a rate of growth of about a thin dime every hundred years. It was very impressive to us. And, although Halmi took many pictures, Dyer was always alert not to let him touch them, or to reach up and bump them. This will be pure beauty that can be seen in the pictures. But it may be many years before this portion of the cave is opened so the tourist can see them.
>
> "This is Burke Miller returning you to WSM."

Jim grinned. He liked to hear people's reactions to these finest gypsum flowers in the world. Buried deep in him was a kind of pride that made him beam all over at others' appreciation of the marvels that could now be seen largely through his earlier exploring efforts.

Charlton and some others took pink plastic plates from the pile and formed a line to be fed. Nancy ladled out a goulash of rice, meatballs, and raisins, garnished with carrot sticks. Gradually the rest of the group joined in, and within a half hour the pot was emptied.

Dyer made up an exploring party headed by Charlton to trace the side sleeping gallery north of the camp. Three men went off, and in about twenty minutes they returned to report to Jim.

"We went into the gallery at that passage just before you get to Fool's Dome . . . had no idea anyone was sleeping there," said Charlton. "They raised a big fuss so we didn't stay long. We moved north in a kind of a stoop passage, about eight feet wide. There were two openings to the Lost Passage. The whole thing seems to parallel it for about four hundred feet where it ends in breakdown."

"Thanks, boys," said Jim.

Some of the gypsum flowers in Floyd's Lost Passage grow to a length of 12 inches, large by comparison to those in the tourist route 200 feet above them. *Photo by James Dyer.*

A supply team rolled in and some of those who had been sleeping paused to watch the always novel procedure of unloading the bulging Gurnee cans. A huge quantity of stores would fit in provided they were packed carefully, and the unloading process reminded us of the circus trick in which thirty midgets emerge from a tiny automobile.

About 3:30 in the afternoon the climbing team assembled in front of Jim Dyer. Lou Lutz clanked into the circle loaded with mountain-climbing hardware: pitons, karabiners, rope, and a piton hammer. Audrey Blakesley wore rubber-soled sneakers, and looked surprisingly pretty, in spite of having been awake for many hours working in the kitchen. Earl Thierry had been assigned to the group. He stood several inches over six feet; he was big and muscular. He looked as though you could climb up him and, by standing on his shoulders, reach the top of any dome. Halmi was busy in the corner assembling his electronic flash gear. He asked if I could get away to hold his extension flash because he wanted to be sure to get well-lighted pictures of whatever they were going to do. I asked Jim if he could spare me from the kitchen where I'd been working for the last eighteen hours. He said he could.

"Is everyone ready?" asked Jim. He didn't wait for an answer. "We want you to see if you can get up into a hole in the top of Fool's Dome. For years now we've been looking for some way to reach a passage we think lies directly over the Lost Passage, but we've never had the equipment or rock-climbing know-how to do it. Thierry knows the way."

We tramped off, leaving the lights of Camp One behind. Thierry stopped in the center of Lost Passage about seven hundred feet from Camp One. He pointed to an alcove at the side of the passage. "There it is," he said.

I lingered behind with Halmi as the others went in; Skeets Miller and Phil Harsham joined us. We entered the alcove to see, over our heads, a dome going up about forty feet from the floor; at the top, a small hole about eighteen inches in diameter led upward into blackness. The dome looked deceptively easy to climb. We could see handholds part way up, but then the wall changed into a steeply sloping flowstone slide. Above this, it looked as though the going became easy again until the climber might reach what looked like a comfortable ledge. By the looks of it, one could stand neatly on the ledge and grasp the edge of the hole with both hands, pulling himself into it. This last maneuver would be tricky, but a clever rock climber could do it. Lou would try first. He ran his experienced eye over the wall inch by inch. I couldn't help thinking as I looked at him that this wiry little man, only five feet four inches tall, doubtless made up in nerve what he lacked in size. He wore a white mustache fringed with a two-day growth of white stubble. I hoped I'd be that active when I was in my fifties.

He was going up now, having picked a spot on the right wall where a crevice led into a passage back of the dome. He tested each foothold carefully, jamming his spiked climbing boots into the cracks. His fingers fished for holds along projecting rocks. He was about five feet up now. He began to inch his way to the left to get into position for an assault at the flowstone slide some two feet above him. Then he stopped.

"I've run out of footholds," he said.

"Stand on my shoulder," said Earl moving into position under him. Earl braced himself against the wall as the tricuni-nailed boot sank into his shoulder. Lou moved slightly, but again he stopped.

"It's no use . . . I can't reach any more handholds." Lou began a slow retreat, ending when he reached the bottom. "Maybe Audrey can do it. She's got a longer reach," he said. Audrey picked her way up the wall like a veteran, then moved over to stand on Earl's shoulder. She encountered the same difficulty, but Earl suggested a move sideways, perhaps giving her a chance at some handholds farther along the wall to the left. We held our breath as Earl slowly shifted his weight to his left foot, dragged his right closer to him, and moved. Audrey met with better luck this time. She found two handholds and was able to take her weight off Earl.

Getting up, however, was another matter. The holds were too high up in the sheer wall for her to get the needed leverage. She returned to the ground.

Skeets Miller whispered to me, "Do you think they can climb it?"

"I think so," I answered. "There's always a way up. It just takes a while to find it sometimes." I had been studying a possible alternate way up, so I asked Earl to consider it on his attempt. The plan of attack lay in going up the crevice, as the others had done, but to a point some two feet higher. There appeared to be a series of handholds here that the others had overlooked. Earl was skeptical but agreed to try.

He reached the point without trouble, extended his left arm, and started a side traverse to the left. He was about eight feet above the boulder-strewn floor, hugging the wall with his body spread-eagled between four holds. "This is just fine!" he yelled. "But I can't move!"

"Can you come back?" I asked.

"Not without losing my balance!" A fall here could be a serious thing. I chimneyed up the crevice about three feet, and climbed the rest of the way up to him. Lou stood below in position to break his fall if he should slip. I stopped at a place where I could wedge my body against a projecting boulder at my back, and the wall at my feet. I put one hand firmly against his foot, pressing it against the wall.

"Good, I think I can make it now." He moved his body around, coming to rest in my lap. We untangled and descended to the floor of the dome.

The party was almost ready to give up, but Lou made a last-minute reconnaissance of the wall for cracks capable of taking a piton. There were none.

By this time I had evolved an elaborate method of climbing to the top, theoretically, at least. I started to explain it, but was stopped by a chorus of voices telling me to try it rather than explain it. I was a bit apprehensive, never having done rock climbing before in the mountaineering sense of the word.

My proposed route lay up the same crevice that Earl and the others had chosen, but the traverse to the left was to be made at a point between that tried by Thierry and the point used by Audrey and Lou. I had moved to

Lutz tries to climb up from his perch on Thierry's shoulders, but he can't find handholds. *Photo by Robert Halmi.*

the left about a foot farther than Earl had when I reached my first crisis. The wall just didn't have any more handholds! My left hand raced over the surface of the wall, feeling every possible chink in the rock, every protrusion. My right hand was growing weak. I moved my hand up to the very limit of my reach, where I found a splendid hole two knuckles deep. I was on my way, I felt.

Guided by directions shouted from below I swung my feet to a handhold Lou had utilized on his first try. There was room for the toes of both shoes on the ledge, but the handhold high in the wall dictated that I remain crushed flat against it. No chance to look for handholds from here on. The only thing I could depend on was my sense of touch. The blood was beginning to leave my arms now. I moved the left arm out, covering the wall with exploratory circles. My arm was very weak now. There wasn't any hold! No way to go but down! Automatically I continued the search with my fingers. Below, the murmur of the group had stilled. Someone, in

Roger   Brucker   starts
up   at   Fool's   Dome.
Photo by Robert Halmi.

a low voice, asked Earl to move under me. They could see my plight. Ten feet in the air and not a chance. Emotions play strange tricks on you when fear takes over. I snapped to with a realization that I'd been a fool to gamble on one handhold when there could be others. I summoned all the strength I could, lifting my left arm to the handhold where my right hand was hanging on. Grasping it firmly, I lowered my right arm and rested it. I felt better.

After a few moments my right arm encountered another handhold slightly below the first. Out of the corner of my eye I could see the V-shaped crack marking the lower edge of the flowstone slide. I could almost reach it. The new hold allowed me an extra two inches' reach to the left. It proved to be an excellent handhold, one I could depend on with a feeling of security, something few other handholds on the wall possessed. If I could just get my left knee into the bottom of the V, I'd be all set to move up! I closed my eyes a moment, then launched out with my left leg aiming blindly at the V. I missed! I was slipping. "Catch me!" I yelled. My left hand locked on the handhold. I felt as if it were crushing the rock into a damp powder. I lurched and stopped.

My left leg had caught and held in a foothold I couldn't see, one which was invisible to those below; it was the key foothold. I put my weight on it, finding it reassuring, comfortable. The rest would be easy. I breathed a deep breath, then another. Skeets was yelling from below, "Come down . . . come down . . . it isn't worth it!" But I had reached the point of no return. Two choices were open. One, let myself go, depending on someone below to break my fall and perhaps a back with it, or two, go on up. From where I clung, it was easier to go up.

I inhaled deeply, feeling a new surge of strength. I pulled and pushed myself up, moving out of the squatting position in which I had landed. I was flat on my stomach on the flowstone slide and its cool dampness felt wonderfully refreshing. Friction from my coveralls prevented me from slipping back. Handholds were more numerous now.

Two thirds of the way up my light went out. I called for those below to light my way to the ledge. In a few seconds I climbed over the lip of the ledge to sit down. Far from being the comfortable ledge we had imagined it to be from below, the top of it sloped about 20 degrees into the yawning space below, and it was covered by a slick coating of mud! I sat on it anyway, knowing I'd have to sit quietly without moving if I intended to get back in one piece.

Someone wanted to know how I intended to get down. I wondered too, but I was too exhausted to think about it for the time being. I requested ten minutes to rest before committing myself to any course of action. As I sat there the hole was just a few feet over my head, and even without my light it looked as inviting as it did from below. But I was in no shape to tackle it.

An eternity passed. My heartbeat settled down to about twice normal and the perspiration stopped pouring from my skin. I made plans to get down.

"Anyone have any string down there?" I yelled.

"I've got some," shouted Earl. "I'll tie a rock on one end and heave it up."

"I warn you, you'll have to get it to me. I'm in no position to reach for it!" I explained the ledge situation to them and they seemed sympathetic.

Thierry tries to climb to Brucker's aid, while Lou Lutz stands by below. *Photo by Robert Halmi.*

The group watches anxiously as Brucker pulls up a fresh carbide lamp. Halmi readies his camera for shot. *Photo by John Spence.*

Brucker smokes a cigarette as he rests after climbing Fool's Dome. *Photo by Roger McClure.*

On the fourth try the rock landed neatly in my lap with the string tangled up around it. It took me ten minutes to get the string unwound and the rock lowered back. Soon the rope was on its way, a new nylon rope, which I pulled up until I held the half-way point firmly in my hands, then threw down the other end. A fresh carbide lamp was tied on and I pulled it up. I was back in business.

The next job was to rig up a place where the rope could be secured. Some projecting fingers of limestone on the wall above me and to the right proved out of reach. Then I remembered the way Lou jingled!

Up came the pitons, karabiner, and hammer. There were all kinds of pitons for any use, much like the ones I had seen before in equipment catalogs. They ranged from three to six inches in length and were made of steel. The pointed end varied for the type of crevice in which it was supposed to be driven. A vertical crevice requires a flat piton with a blade in the same plane as the eye, while a piton for a horizontal crack has a 90-degree twist between blade and eye. I chose a piton with a horizontal blade and edged around on the ledge slowly, planting my feet firmly with each move. Directly under the hole I found a likely spot for driving the piton.

"Remember," shouted Lou, "it's not safe until you hear it 'ping'!" I pounded vigorously on the piton with no sign at all of a pinging sound. It sank slowly in, then plunged in beyond the hole in the end! Then I discovered that the wall in which I had been pounding was covered by flowstone and backed by mud. The entire wall was the same way. I did the best thing I could under the circumstances; I picked a place where the flowstone looked especially thick and began to hammer another piton in behind it. By the time I had driven it most of the way in, it was making a metallic "Tank, tank" sound. I hooked the karabiner in the eye of the piton, ran the rope through the karabiner, and called for a test from below. Four people hung on the rope. It held.

The way was now clear for Lou to come up. He tied a bowline around his waist while two persons made ready to belay him. He climbed up with no trouble and soon sat on the slippery ledge beside me. "We were a little worried there for a while," he said.

"So was I," I admitted.

"Now let's take a look at that hole. Will you give me a boost?" I stood up cautiously. Lou grasped the sides of the hole and I pushed him upward. There was a pile of loose rubble at the lip of the hole which he cleared away after giving warnings to those below. One particular block, as big as a typewriter case, tilted downward into the hole, offering him a secure handhold. It apparently had been cemented in this position by flowstone.

He was still belayed by the piton seven feet below him. Both arms went over his head, the left forearm curling around the boulder. He went swinging out, giving us an anxious moment, but he had aimed for a high head buttress on the opposite side of the hole. From here he pulled himself in and disappeared from view.

"I'm in a small circular chamber about eight feet in diameter . . . ceiling about two to three feet. There's one narrow fissure . . . let's see . . . it leads off about ten feet and ends in breakdown. There are two more cracks, too small for a man, but they're both filled anyway."

I relayed the word back to those waiting below. Lou came out of the hole covered with dust and perspiring freely. It was a shame to go to all that struggle and effort only to find a blind lead. But that's what cave exploring is. You check every possibility. Then the ones that don't lead to more cave can be checked off the list, allowing more time to devote to the others. It's the x factor that keeps cave explorers going, and every time you can eliminate one unknown x from the cave, you've done something worthwhile in terms of exploration. We won't have to go back to Fool's Dome again, but if we hadn't conquered it, some men would lie awake nights wondering what lay beyond the hole in the roof of the dome. Not many men, perhaps, but I would. Now we knew.

The descent with the aid of rope was easy. Lou followed me. A mixture of disappointment and triumph was reflected in the faces of all those who stood at the bottom. There was triumph in the fact that working together we had mastered the dome; we had reached our objective, and we all had formed deep and lasting friendships, based on respect and mutual admiration of each other's efforts. We eased our disappointment by going for supper. Expedition Leader Lawrence had just arrived in Camp One after a short trip with Luther Miller. I asked him if he had any news for us about his activities since entering the cave.

Upon his arrival at Camp One at 7:45 P.M. on Monday, he found there was a shortage of sleeping bags; or rather, a shortage was created when too many people wanted to sleep at once, while at other times many bags were idle. Joe had tried (without complete success) to get people better distributed on three sleeping shifts, setting up a system whereby each person would place a tag at the foot of his sleeping bag bearing his name and the date and time he crawled in. There were still several people in camp who did not have sleeping bags, so Joe lumped them together in a party under Earl Thierry and sent them out to explore in the level above Floyd's Lost Passage.

Earlier Luther Miller had stood by impatiently pawing the earth. He asked Joe if he wanted to see "some cave."

"Yes, I'd like to see some of the known passages so that I can better orient myself," answered Joe. Luther and Joe signed out in the log, then walked down to Mud Avenue. Luther took off like a hound after a rabbit. Lawrence puffed along behind as fast as he could move. They sped along a dirty-brown canyon to a room Luther called Five Passages, where no fewer than seven passages led out. They took one that led to a crawl going in both directions. Luther dived in and started crawling through a smooth limestone tube, with Joe following him over the clean sand floor. The sand got under his knee-pads and cut into his knees.

After going about twenty feet, Luther said, "This doesn't look familiar. You wait here and I'll see what's up ahead."

Joe lay in the sand completely relaxed and at peace with the world. It was quiet, cool, and comfortable here. What more could he ask? Judging by the way he spoke, he had a feeling bordering on love for this limestone wrapped about him.

Soon Luther was back to report that they had taken a wrong turn. They should have gone the other way in this crawlway, and Joe complained good-naturedly as they backed through a crawlway too small to turn around in.

Once straightened out, they gained momentum as they dashed to the top of the X Pit, swept in a wide circle over Floyd's Jump-Off and down the C Trail, and finally back to Camp One without retracing their steps except for the short confusion in the limestone tube. They had covered part of the cave on the entrance side of Camp One. Joe had gained a feel for the layout of the passages. And he wore the carefree smile of a man who felt "at home" with the world around and above him.

By Monday evening Camp One is a going concern. Hungry explorers and support groups are fed around the clock from the kitchen in Floyd's Lost Passage. *Photo by Charles Kacsur.*

12

ASSAULT ON B TRAIL

EXPEDITION LEADER JOE LAWRENCE planned to devote a major effort in the B Trail area at the start of the expedition. During the three-day trip to the lower levels in 1951 Charlton and Austin had entered it and had taken a side route known as Bogardus Waterfall Trail. At their utmost point of penetration, they reported they had not begun to exploit its potential. This reconnaissance shaped Joe's decision to launch the assault on B Trail as early as possible in the week, starting with the advance party's search for a possible site for a second camp.

But why pick B Trail in particular out of the scores of other passages, any one of which might keep the expedition busy all week? For one thing, B Trail represented the point of farthest penetration into the cave. It held possibilities we knew about. It was heading into the heart of Flint Ridge, where men had long suspected the presence of caves that would dwarf even this underground system.

Although the advance party had not found a site for a second camp, they had verified the potential that lay there, and for this reason Joe had ordered a survey to be started and had directed Dyer to send teams of explorers ahead of them to scout the way. Jack Lehrberger and Luther Miller played a leading role in this exploration. They were inexhaustible; it seemed that neither of them required sleep. Parties attached to them complained that they never required rest stops and that the pace they set in scrambling over rocks called for the most grueling speed possible.

Acting on information obtained from the advance party, they had explored a hole in the ceiling of B Trail about one hundred feet from its apparent end. Chimneying up through breakdown they entered virgin passages lined with gypsum on a higher level. They passed through a series of small rooms, coming finally to a place where they could see into another passage. Lehrberger moved rocks the size of watermelons to open the way, then they continued. In another ten minutes they found an arrow on the rough rock wall, and footprints on the floor. They had come from virgin cave into explored cave, but they didn't know where they were. Excitement spurred them on. Running in a crouched position on a narrow trail, they soon arrived at a small room with initials smoked on the wall. *J.D.*—Jim Dyer. The "old man" had beaten them to it, they thought, but realized that he would have had to enter by a different route, since the passage through which they had come had never before been entered. Looking out a side passage they discovered a chimney leading down with spike marks on the walls. This was the way Dyer had come.

Lehrberger poked his head through the tiny hole leading to the chimney. From his vantage point at the very top of the shaft he estimated its depth at sixty feet. His eyes moved from rock to rock, examining the telltale white scratches of Jim's traditional golf shoes. The vertical tube was about

When you enter virgin cave, you never know what you were going to find. Maybe a big pit, a high room, or maybe nothing. But Luther Miller has to find out. *Photo by Robert Halmi.*

four feet in diameter, with muddy, water-cut walls. Grotesque projections, fragile in appearance yet evidently strong enough to hold the weight of a man, had served as hand and footholds for Jim. Lehrberger's lamp beam picked out the glint of something shiny below. It was a pool of water.

He retreated into the muddy passage at the top, then worked his feet through the hole until his legs and trunk dangled into the pit. With a practiced glance, he chose a small foothold, testing it while still hanging on to make sure it would support his weight. Then he pulled the rest of his body into the shaft. His next foothold was about two feet lower and behind him, and he reached it by carefully extending his left leg backward. Again he repeated the slow, deliberate process of searching for a safe handhold. He realized that the holds would have to be more secure than usual, for his leather gloves were soaked with muddy water from crawling through the passage above. Removing the gloves would not help; experience had taught him that using bare hands on mud covered rocks is an invitation to disaster.

Mountain climbing has its hazards, but none so deceiving as mud-covered rock under dim illumination. On ice, the mountain climber sees and respects the treacherous footholds, but in a dark cave, mud-covered rock is not always so easy to discern. For this reason the cave explorer's life depends on his awareness of the potential dangers of the unseen. These thoughts were in the forefront of Lehrberger's mind as he inched downward toward the water below. His years of cave-exploring experience had sharpened rather than dulled his sense of precaution.

At the bottom he yelled instructions for Luther to follow him, and within five minutes both men stood in a low room looking upward into the chimney. Luther moved around the edge of a shallow pool of water into a passage whose floor bore spike marks. They paused briefly to view with awe a profusion of stalactites, stalagmites and columns. The largest column was about four feet high and almost a foot in diameter. But more interesting to them were the smaller, perfectly formed stalactites. Some were pure white. Others were stained a dull orange, a color peculiar to the limestone in this portion of the cave. Drops of water plunged from the tip ends at intervals, spattering on smaller stalagmites below. Jack suggested they continue their journey.

In another room they found arrows pointing out. Three hours later they emerged from Bogardus Waterfall Trail into B Trail and raced back to camp with the news.

What they had discovered in effect was a link shortening the time from Camp One to Bogardus Waterfall by two hours.

Almost immediately the survey of B Trail was started. Early Monday morning Earl Thierry led a four-man party to the entrance of B Trail. At B Point, an arbitrary starting mark, they slowly and systematically surveyed their way in. After covering 1710 feet and forty-seven survey stations, they reached their endurance limit. At first glance, this appeared to be a ridiculously short distance to an endurance limit. But this operation had taken them six hours and had worn out the party who had been ready for sleep hours prior to the time they had started.

Joe Lawrence conducts a large share of the expedition's business by telephone from Camp One. *Photo by Robert Halmi.*

Bob Handley led a second survey party of four out B Trail to continue the survey from Earl Thierry's last station. Ackie Loyd, who accompanied Handley, had been in the expedition's first exploring party to penetrate B Trail. On his trip, Ackie had pushed farther than anyone before, shoving back the endurance limit. He now guided Handley back over the route of his exploration so that it could be measured and recorded.

From station to station they traveled. They established each station as far from the last as they could and still have it in line of sight. It was a twisting trail, so they had to keep the distances between stations short. Finally they reached a point where they could go no farther, where the ceiling barely cleared the four-inch thickness of mud which covered the floor. Sixty-four people working as a team had made it possible not only to visit, but to measure and record a part of the cave that had previously been beyond reach.

But the penetrable limit of B Trail didn't mean the end of the cave, for Handley's surveyors had passed many side passages. They had only begun; there was much more to explore and survey.

Joe Lawrence had given an assignment to Roy Charlton, Roger McClure and me. We were to push our way out Bogardus Waterfall Trail, surveying it as we went. There was no choice in the matter, since sleeping bags would not be available for another eight hours.

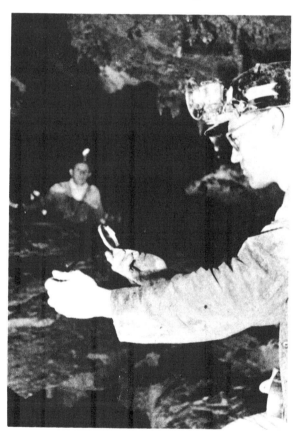

Cave surveyors use a Brunton compass to measure the horizontal and vertical angles between two points. A tape measure will give the distance. These three coordinates define the location of one point with respect to another in three-dimensional space. When measurements are made from point to point in a passage, then plotted on paper, a broken line shows the course of the passage. *Photo by R. J. Richter.*

The Bogardus Waterfall Trail had been discovered on an earlier trip by Bud Bogardus, Bill Austin and Jim Dyer, who had crawled along it for several hours. When they had traveled as far as their time limit would allow, Jim suggested they rest, but Bogardus was like a bloodhound with a scent. He climbed to the top of an unlikely looking chimney and discovered a network of passages. Following what appeared to be the main passage he arrived at the waterfall that now bears his name. There were side leads off Bogardus Waterfall Trail, and they too needed to be explored. A map would tell explorers where the side leads lay and in what direction they headed.

Surveying can be deadly dull, or it can be the most exciting part of cave exploring, depending on the potential uncovered as you progress. I gathered a supply of rations for an eight-hour trip.

We signed out in the Camp One log book at 9:10 P.M. Monday. I carried the field book for recording survey notes. Roy looped the snake-like steel surveying chain over his shoulder, and McClure tucked the precious Brunton compass in his coverall pocket. We moved down the Lost Passage, crossed the yawning entrance to Mud Avenue on a lodged boulder, and stood looking into the entrance to B Trail. The passages formed a "Y" at this point, with B Trail leading off from the left-hand fork, and the Lost Passage forming the other two legs. It was an inviting entrance, about twelve feet high with a width of about ten feet. We moved into it.

The inviting entrance proved to be a deceptive lure, urging the cave explorer on. But the passage rapidly diminished in size to about five feet in height and about nine feet wide. Walls and ceilings glittered under our carbide lamps as the crusted gypsum, pure white, almost like icing on a cake, reflected and re-reflected every ray of light. The floor was piled deeply with broken gypsum, in some places ankle deep. We crouched down and scurried onward for about fifteen minutes, following the tiny survey marks indicating the course of the main passage. The stations slipped by in a blur: B-8, B-9, B-10.

"Bogardus Waterfall Trail takes off to the left, just beyond B-17," yelled Roger, who knew the location from his previous surveying activities. As we moved I realized how important it was that each station be given not only a number, but also a letter designation. In that way, an explorer coming upon the passage from another direction would know immediately which passage he had entered. By checking the map, which would come later, he could easily determine exactly where he was in relation to the rest of the cave, something no one had been able to do up to now.

Station B-17 was marked on a prominent rock in the middle of the meandering passage. Beyond it a small hole led downward to Bogardus Waterfall Trail. We adjusted our lamps and sat down to rest.

In a few minutes Roger was up again. He moved back to B-17 and took out the compass. "We might as well set up the first station and get started," he said. "What letter designation will we use?"

"Let's use *WF*," I suggested. I took out the notebook and began to write. Roy turned up the flame of his carbide lamp by adjusting the water valve. A large tongue of fire shot forth, and with it he placed a soot dot on a rock directly above the hole. Beside it he marked *WF*. McClure lined up WF in the mirror of the compass lid, then glanced down at the needle. "North 83 degrees West," he dictated. I noted this in the proper column in the book. "Now, let's get the vertical," he said. He turned the compass on its side this time, sighting through the cross hairs. With his fingers he manipulated the tiny level inside. "Minus 5 degrees," he said.

Roy grasped the end of the chain and held it on the dot. "Chain," he shouted. This signal indicated he was ready on his end for the distance to be read. McClure squinted at the band of bright steel pulled taut over B-17. The distance was fifteen and a half feet. I marked this in the book.

Roy crawled into the hole and moved down to establish WF-1. With its dot marked on the wall, McClure took the compass reading from WF to WF-1. He read the vertical and the distance. The survey was under way.

Surveying in a cave is a specialized job out of the realm of surveying in the usual sense. A transit and stadia rod are virtually useless because of their size. Merely dragging the transit through a crawlway would make it unreliable, and the weight and bulk of both would present a problem. Further, the observer would find it extremely difficult to use an optical instrument under minimum light conditions. Unless a cave has very large passages throughout, a would-be cave surveyor is reduced to using a small but nevertheless accurate surveying compass. It fits neatly in the palm of the hand and is ruggedly built of aluminum to stand considerable punishment. By means of a mirror sighting arrangement, the user can simultaneously see the station on which he is taking a sight and the compass reading in degrees. Accuracy is further aided by a built-in spirit level which is lined up prior to making the reading. Another level, or clinometer, adjustable and calibrated in degrees, enables the user to determine the vertical displacement of the next station from the horizontal, so distances will be accurate when the map is drawn up from the data.

The whole purpose of drawing a map is to determine how passages lie in relation to one another, and in relation to the surface. Ideally, a complete map of the known portions of a cave would give an explorer vital clues for the discovery of still more cave. For instance, passages leading into a hill or ridge are likely to prove more fruitful than those heading toward a valley. A passage ending in breakdown often can be considered as holding no more possibility for discovery if the map indicates that the breakdown is directly under a surface sinkhole.

Bogardus Waterfall Trail was now trending southwest, toward the center of Flint Ridge, perhaps toward big cave. McClure's sights covered twenty or thirty feet now. The passage was meandering, like a mature stream, but its general direction continued constant. Hour after hour we surveyed along this passage. It had changed little in character from the time we had entered it. Jagged gray rocks littered the floor and the ceiling was just high enough to permit us to crawl comfortably. The routine continued: crawl awhile, stop, make notes, then crawl some more.

I was terribly tired. Moving after making notes became an agonizing job. We stopped and rested. I stretched out on my back and looked at the ceiling, running my burning eyes over large veins of black flint criss-crossing the limestone. On my left, some of the flint was shiny gray, and some was dull orange. Indians would have had a field day here, I thought.

"Wait!" a voice shouted a long way back on the trail by which we had come. I looked at Roy who was looking at McClure. Then we all stared in the direction of the call.

"Wonder who that is?" said Roy.

"I don't know, but we'd better wait," suggested Roger. We lay out flat under the low irregular ceiling. Finding a comfortable spot was difficult among the breakdown, but we piled some of the more irritating rocks out of the way.

After ten minutes of waiting we heard no further sound, neither voice, nor the tell-tale racket of a man scrambling toward us.

"We *did* hear someone yell, didn't we?" asked Roger.

"I plainly heard the word 'wait,'" I said.

"Could be that some explorer is looking out side leads on B Trail. He probably yelled at the rest of his party," offered Roy.

"Then maybe it's the 'little men,'" said Roger with a broad grin.

The "little men" was the term often applied by cavers to explain the origin of certain inexplicable noises. Cave explorers are not superstitious by nature, for if they were, they probably would not enter caves in the first place. Yet the "little men" had "spoken" before to several members of the expedition. Bill Austin and Jim Dyer had entered the cave one night and were walking along the commercial route when they heard a voice calling clearly to them: "Wait for me." They stopped, but no one came. To their knowledge, no one else was in the cave, for hadn't they locked the door behind them when they entered?

The expedition leader himself had related a similar experience to us shortly before we entered the cave. On a three-day trip, he and another man had waited for a half hour for a party to catch up with them at Ebb and Flow Falls. They had thought others were coming because they had heard the babble of voices which usually heralds the approach of another party. No one had come. We resolved to check the log on our return for records of other parties in the area. Eyelids drooped as the work continued. Voices were droning at me; I was falling asleep. I shook my head and continued crawling. A comparatively large side passage departed from the left-hand wall ahead, so I volunteered to explore it in an attempt to keep awake. In five minutes I emerged in a bigger passage: New cave! I thought. From my right came a voice, "Find anything?" My discovery was just another cut-around leading into the main passage. I sketched it in on the rough passage map I had been keeping beside the survey notes.

We stopped to eat at the edge of a snaking canyon cutting across the passage at right angles. A draft of air sprang up, coming from the passage ahead of us. We looked at each other; big cave? The breeze was a good sign of it.

"Feel that draft!" said Roy, munching on a chocolate bar.

"I think we're on the trail of something big," said McClure. We continued to eat in silence.

Somewhere ahead of us was an orifice, perhaps leading to another cave, or perhaps to the outside world. Air was spilling through, probably due to an outside pressure change. There was always the likelihood, however, that the orifice would be too small to go through.

We started into the forty series of survey stations. According to an earlier tally, we had come about fifteen hundred feet. The readings were generally southwest now. The breeze blew for a few moments, then subsided, and started again. We had left the sewer-type passage with its characteristic tubular walls, and we were now in an area which had been cut by streams. Canyons meandered across the floor of the passage. The walls were a deep brown color. I sat down to wait for the next reading.

"Did you get that?" It was McClure's voice. I opened my eyes. "Did you get that?" he repeated.

"Get what?" I asked.

"That last vertical reading?"

"Vertical reading? I didn't even get the compass reading!" I knew I could not control my fatigue. He repeated the readings for me. We surveyed two more stations, up to WF-57, when I announced that I couldn't go any farther. Roy was yelling something about being almost at Bogardus Waterfall, but it couldn't have interested me less. I crawled after them, almost in a stupor. My eyes would not stay open. We were crawling on a sandy floor now. I stretched out.

"I'm going to get some sleep," I yelled to Roy.

"You can't sleep here," he said. "You'll get too cold."

I fished out my sweatshirt and dangled it drunkenly over my head. "I won't get cold, I'll put this on."

"O.K.," said McClure. "Roy says he's going to join you. I'm going to see what these canyons in the floor are all about . . ." his voice trailed off in my consciousness.

I woke up after what seemed to be a split second of wonderful sleep. By my watch, it had been for forty-five minutes. A dull glow shined up from a crack in the floor of the passage near my feet. In a few minutes Rog McClure appeared and came toward me. I was aware of a clattering sound at that instant, as though something had fallen into the crack and had come to rest some distance below.

"The compass!" yelled McClure.

"What about it?" I said.

"I dropped it down the crack."

"You'd better get it," I said, still reclining.

"The crack is too small . . . it went too far down."

"Yell if you need help," I said. I rolled over and went to sleep again.

A half hour later he returned with a downcast look.

"It's out of reach," he said. "I tried to get down into the canyon farther back, but I couldn't. It's too tight."

I got up from my cold sand bed and went over to the narrow crack. I peered down into it, much like a motorist looking under the hood of a stalled car along the side of a road with no idea what to do. The compass was nowhere to be seen and I couldn't care less; fatigue dictated that. Actually, a very good fifty-dollar compass rests at the bottom of a remote canyon in Crystal Cave. Anyone who finds it is welcome to it.

Roy Charlton was back now with the news that he'd found the chimney Bogardus climbed up when he discovered his waterfall. We moved along the passage and soon heard the sound of falling water. We crawled through a small hole and found a pool about six feet wide and about ten feet long. Here and there brown lily-pad formations spread themselves out on the surface of the water. They looked thin and fragile, though made of solid calcium carbonate deposited by water evaporating from the pool. At the edge of the pool were rim-stone dams, also made of the same mineral pre-

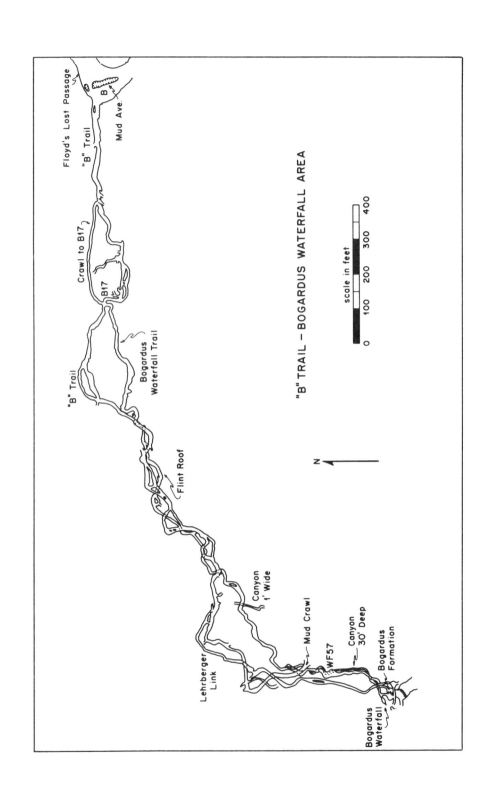

Floyd's Lost Passage

"B" Trail

Mud Ave.

B

Crawl to B17

B17

"B" Trail

Bogardus
Waterfall Trail

Flint Roof

Lehrberger
Link

Canyon
1' Wide

Mud Crawl

WF57

Canyon
30' Deep

Bogardus
Formation

Bogardus
Waterfall

N

"B" TRAIL – BOGARDUS WATERFALL AREA

scale in feet

0   100   200   300   400

cipitated out from the water in the pool. Wind fairly whistled out from under a ledge. Back under it was a muddy-looking crawlway, hardly big enough to enter, but we were in no condition, physically or mentally, to explore it. Instead, we followed Roy up the rather hazardous climb to the top of the formation.

Sixty feet above Bogardus Waterfall Trail we slithered through a muddy tube into a low room. Initials on the wall indicated that Jack Lehrberger, Luther Miller, and Jim Dyer had been there before. An arrow on the wall pointed the way out. We stopped to eat the rest of our food. I was feeling a little less tired now.

"I suppose we ought to start back," said Roy. "We've been gone for nine hours." We agreed, and began to move slowly along the passage. We were astounded to see the footprints of only two people along the passage. This must have been the new link Jack and Luther had found. In twenty minutes we climbed through a hole in the floor at a spot comfortingly marked B-53. The trip back would be easier than the trip in, Roy assured us.

It was 9 o'clock in the morning when we stumbled into Camp One. I turned the valuable survey notes over to Earl Thierry who would phone them to the surface for plotting.

All three of us converged on the log to discover the identity of the "little men." Now we felt again the eerie feeling we had felt before. No one had entered B Trail during the time we had been gone!

I ate a plateful of Spanish rice, then mechanically shuffled down the passage to a sleeping bag. Remembering Halmi's unhappy experience with damp clothes, I folded my coveralls and placed them between the bag and the air mattress hoping that my body heat would dry them out. I crawled inside to spend my first "night" sleeping in a cave. Mentally I calculated the hours making up this underground activity of mine. Fifty-three hours was a long time without real sleep.

When the survey notes from B Trail and Bogardus Waterfall Trail were plotted, all of us were amazed to learn that both of them lay directly under the commercial route for part of their extent. Common sense had told us that Bogardus Waterfall Trail and B Trail were two prongs of a large "Y." Now we saw that Bogardus actually lay under B Trail, crossing and recrossing under it no less than ten times.

As exploring parties in this area continued to report back on their penetrations, it became evident that the most promising leads had been explored. Joe Lawrence felt that labors in the B Trail and Bogardus Waterfall area were giving the expedition diminishing returns. For this reason, he turned our major efforts for the rest of the week toward other areas of the cave.

13

MEN WITH BOTTLES

CONCURRENT WITH OUR OPERATIONS in the B Trail-Bogardus area, the expedition's scientific staff worked on the systematic study of Crystal Cave's underground environment. Barr and Richter had already filled several dozen soil-sample bottles with earths from various parts of the cave. It was their aim to fill as many of the six hundred aluminum bottles as they could, and to achieve this they gave six to every outgoing party. Each party member was instructed to collect a soil sample whenever he found himself in a new part of the cave. In this way, bottles began pouring back to the scientific stockpile located on a ledge behind Camp One kitchen. Some of the bottles rattled. Some were heavier than others, depending on whether they contained mud, dust, gravel, or molds found on decaying organic matter.

At the end of the expedition Richter would deliver the bottles to a large pharmaceutical laboratory in Brooklyn, New York. There they would be added to a collection of samples gathered from soils all over the world by just such expeditions as this, travelers, missionaries, and members of the armed forces. The whole purpose of this extensive program was to provide raw material for scientists looking for new varieties of mold from which to extract antibiotics. Terramycin had been discovered as a direct result of such efforts, and scientists believed that others could be isolated in the same way that penicillin and streptomycin had been selected from the many varieties of molds under study. When our collecting was completed, the laboratory would take over.

Most soils contain molds of some variety. It would be the technicians'

Explorers collect samples of soil in the cave in aluminum bottles. The bottles are tightly closed and sent to a laboratory so their contents can be examined for molds that may yield new antibiotics. *Photo by R. J. Richter.*

job to isolate these for study. The process has been developed to a standard routine, starting when each sample is diluted with sterile water so that each individual drop contains only a few mold spores. One drop is smeared over a nutritive substance having the consistency of firm jelly; in it are carbohydrates, proteins, and minerals necessary for mold growth. Several drops from each of the many soils are smeared on the nutritive material in little glass plates called Petri dishes. The dishes are placed in a warm room for several days so each mold can grow and form a colony, the latter looking much like the spots of mold seen on stale bread. From their different sizes, colors, and shapes, scientists identify them and single out those they wish to study.

Molds which they select are placed in a test tube under sterile conditions maintained to permit the complete isolation of the variety under study. With more food, the pure culture continues growing, until it is placed in a liquid medium in which the secretions of the mold will accumulate. The liquid is then tested for the presence of an antibiotic by adding a few drops of it to a pure culture of human disease bacteria. If the bacteria cease growing, it signifies the presence of an antibiotic. Other tests determine whether the antibiotic is one already known, or a new one.

Assuming a new antibiotic is found, scientists grow it extensively in hundreds of gallons of the liquid medium to get sample quantities for testing in egg embryos infected with a wide variety of germs. If the antibiotic

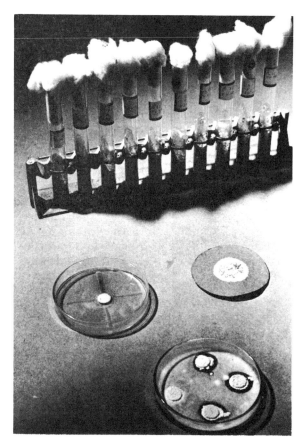

As in the case of Terramycin, a potential antibiotic is tested against Petri dishes full of germs (foreground) or against streaks of different bacteria (left). A translucent area in the germ colony means that the new drug has killed the pathogens growing in that area. *Photo by Chas. Pfizer & Company, Inc.*

passes these tests (and very few of them do) it is further tested for possible harmful effects on animals, and later, on humans. Doctors then test samples selected in the laboratory on patients who fail to respond to other treatment. If they find the antibiotic effective, a new wonder drug is announced.

The outcome of the test program on soil samples gathered from Crystal Cave would not be known for years, but the fact that these had been collected in a relatively inaccessible location might open the underground environment to new vistas of antibiotic research.

Animal life had been seen by the advance party. Charlton's carbide light had been reflected by the beady eyes of a pack rat when the party had gone to Floyd's Kitchen to retrieve the two sleeping bags brought in during Doc Wanger's inspection tour. Already the enterprising rodent had attacked the corner of one bag, instinctively knowing that the padding would make good nest material. Some of the other explorers had seen isolated bats clinging to the ceiling. They were varieties found hibernating in most caves. And cave crickets perched on the walls and ceilings almost everywhere.

But Barr's penetrating search for life was keener than ours. As a biologist, he had studied Tennessee caves extensively, and spoke with authority about animal life we had not noticed in Crystal Cave.

In the dual role of expedition biologist and geologist, he was first called into action to identify a curious formation one member had discovered on a ledge along the Lost Passage. Some tiny nodules, hard crusted on the outside, revealed that their inside was composed of a chalk-like liquid. He identified it as "moon-milk," a romantic name given by European speleologists to this precipitate of calcium carbonate. Its discovery in Crystal Cave is one of the first recorded instances of its being found in the United States.

Barr, assisted by Richter, crawled and collected throughout his stay in the cave. Everytime the biologists reappeared with their glass bottles, curious explorers gathered to peer at them and to quiz the biologists about their findings. They patiently answered the questions, then set off in search of more bugs and worms and crickets.

The techniques for finding subterranean animal life were well known to the biologists. They found beetles by searching sand flats near the walls of passages. Other life could be found by turning over one rock after another until a likely specimen turned up. At the bottom of X-Pit, the hunters had stared for hours hoping to get a glimpse of a blind cave fish, and, in another location, Barr had discovered a tiny eyeless spider by concentrating his search on a single piece of rotting wood.

Cave crickets scurry over walls and ceilings but are not quick enough to elude men with bottles. *Photo by Robert Halmi.*

Cournoyer, the meteorologist, suffered a serious set-back in pursuing his studies of the cave's weather. As a participant in the Fourteen-hour Nightmare, he had been forced to abandon his Gurnee can full of meteorological instruments. When they arrived later in the week, he discovered to his horror that, in spite of the clothing packed around his barograph, it had been smashed beyond repair. The thermograph fared somewhat better, however, and he was able to repair it. It occupied a hallowed position in the Lost Passage some distance from Camp One. Everyone approaching it was warned by bold signs to keep his distance to prevent the delicate instrument from recording a sudden rise in temperature due to the body heat of the passerby.

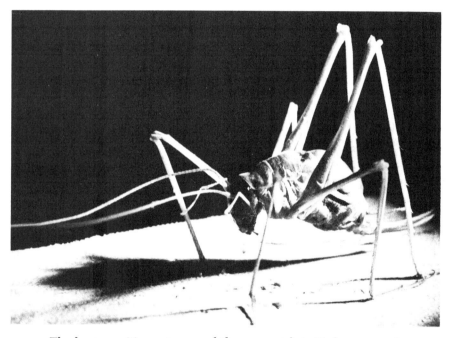

The long, sensitive antennae of the cave cricket, *Hadenoecus subterraneous,* permit it to crawl about safely in total darkness, but the graceful insects do not hesitate to jump off into space, generally landing without damage. *Photo by Charles E. Mohr.*

At regular intervals he removed a sling psychrometer from its case and twirled it in the air. By comparing the wet and dry temperatures of the instrument he could calculate the relative humidity for Camp One. His measurements revealed that humidity in this area normally remained a constant 97 percent and the temperature seemed constant at a cool 54° F.

Meteorologist Cournoyer ran head on into some problems he hadn't anticipated when he signed up for the expedition. Since all of his speleo-meteorological research up to this time had been conducted in small caves, he could not draw on his own experience in dealing with a cave system extending for miles. Published reports gave him no help since practically all subterranean meteorological studies had been confined to small caves or mines, usually having one or two major openings to the surface.

He became acutely aware of this problem one night, when the people around him in Camp One began to complain that it was getting colder. A candle flame, formerly burning normally, was now bent over at a right angle. A breeze was blowing toward the north end of the Lost Passage. The meteorologist checked his thermometer. In less than an hour the temperature had dropped 9 degrees. He found that the humidity was now lowered to 78 percent. He quickly phoned Base Camp and was informed that a storm was brewing outside.

From this evidence, he reasoned that falling air pressure above ground was lowering the air pressure inside the cave and that the flow of air to the north indicated the presence of an orifice in that area. But in this extensive cave system, he couldn't be sure whether the orifice led to the outside, or perhaps to another cave beyond having a surface connection.

Acting on information from Base Camp that rain was beginning to fall outside, Cournoyer set up a staff gage to measure any increase in water level in the Camp One spring. Three hours after the rain had started, the pool at

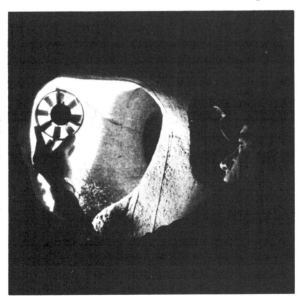

The meteorologist uses an anemometer to measure wind velocity through the twisting passageways of Crystal Cave. He observed some interesting changes in wind direction in Floyd's Lost Passage. *Photo by Robert Halmi.*

the bottom of the spring increased three quarters of an inch in depth, showing that rain water was working its way down through underground solution channels to augment the normal flow of surface drainage.

Cournoyer stated at the end of his studies that possibly the most significant thing he had learned was that cavern systems, apart from individual caves, have many complex factors affecting an analysis of their meteorological phenomena. For instance, it has been generally accepted by many people that temperatures beyond the entrance to a given cave usually remain reasonably constant, and seem to be determined by the average temperature of the region in which the cave is located. Many observers had corroborated these findings by independent studies on widely scattered small caves. But in Crystal Cave, the principle didn't hold. The average mean temperature for the area immediately surrounding Mammoth Cave National Park was determined from United States Weather Bureau reports to be 58.8° F. The average temperature in the Lost Passage during the week was 54.3 degrees. But the meteorologist hastened to state that final conclusions could not be drawn from this because of the relatively short duration of the test observation. Nevertheless, his findings would provide a valuable reconnaissance into the field of big-cave weather study.

Members of the press never tired of questioning the scientific staff regarding their activities and findings. One reporter, not satisfied with communiques issued by the information officer, phoned to Camp One to interview Barr personally. He seemed to feel that the scientists were hiding something, since no announcement had yet been made of "weird creatures of the depths" and "revolutionary new theories." Barr talked with him patiently for almost a half hour and in general gave the man a clear but rather unsensational picture of what the group had found:

REPORTER: I wonder if you can tell me something about the animal life you are finding down there?

BARR: Yes. Shall I give you a general breakdown?

REPORTER: I wish you would.

BARR: There are, of course, three general groups of animals found in caves—one is an especially adapted group which have no eyes, but instead, long antennae or feelers. They are frequently colorless or nearly so. The principal forms of this type that we have studied so far are obligatory cave dwellers such as the cave crickets, the cave beetles, and those blind flatworms we found in one of the streams off Floyd's Passage.

REPORTER: Blind flatworms?

BARR: Yes. Now the crickets and beetles—the crickets have no eyes or extremely reduced eyes—the crickets and beetles are apparently found all through the cave. That is quite interesting because it raises the question of where their food supply comes from and where they breed. The crickets, of course, have no real larval stages.

REPORTER: Beetles do, however, don't they?

BARR: Yes. However, we have been unsuccessful so far in finding either larvae or eggs of these beetles. Possibly we may before the week is out, but already we have made a rather extensive search, without success. It is quite a puzzle as to where they do lay their eggs and where the larvae live in Crystal Cave.

REPORTER: Or what they live on?

BARR: Yes, because crickets are found all through the cave, in places where there seems to be no possible chance of organic food getting to them. Somehow they must find food. The flatworms are very much like the typical planarians that you might see in surface streams. But down here these little flatworms have eye spots that are very much reduced, and they are colorless. That, of course, gives them a white appearance. There are several streams running in and around Floyd's Passage here, small streams. But the planarians are found only in one particular stream. We have developed some theories about why they are found only in this one stream, but I don't think we will come up with anything really definite.

REPORTER: I see.

BARR: Those three forms are the most common of the obligatory caverniculous animals. Then we go on to those kinds that are found in caves but are not necessarily blind, or restricted to them. These include the cave rats and bats. These are the most common members of this group we have run into. There are several nests of cave rats in various places in the cave and a few of our explorers have seen one or two rats. There are not as many cave rats here as we have found in some caves of this nature. Apparently we are rather far away from any sort of opening to the surface, at least where the rats could get in, but still there *are* a few rats around.

REPORTER: They have to come out on the surface?

BARR: Yes. They are true pack rats, sometimes called cave rats or wood rats, and their presence usually indicates some connection with the surface.

REPORTER: Do they store up stuff in there?

BARR: Oh yes, they go in for that sort of thing.

REPORTER: They bring in leaves, nuts, and twigs?

BARR: They pile them all around the nest. But we haven't seen any green leaves here in Crystal. In many caves they do bring in grass and twigs. This time of the year the type of leaves on which rats chew may not be available. We saw one very interesting nest made out of the fibers of rope some earlier explorer had dropped along the way. The fibers were all unraveled and coiled neatly back in the nest. These rats will also pick up flashbulbs, cigarette packages or anything else that catches their fancy. Fortunately, we have not been plagued by them as we have in some other caves. They will steal almost anything they can carry.

Early this morning from about 4 to 8 o'clock Bob Richter and I crawled down the B Trail, which is a stoop, or actually an out-and-out crawlway for more than a quarter of a mile. Near the end it apparently runs close to the surface and there is dripstone. It is damp and quite muddy. We got rather wet in crawling out this passage, but we found in a little pool, surrounded by mud and flowstone, the skeleton of what appears to be a cave rat. Exactly how old it is, is of course conjectural. Anywhere from 1000 to 5 years old. Unfortunately, the jaw bones and teeth were not preserved. The teeth are necessary for absolute identification of the species.

REPORTER: That was in a little stream or pool?

BARR: In a little pool surrounded by stalactites, a limestone pool with rim-stone edges. And then finally the bats. In the commercial portion of the cave, I found the largest cave species in this part of the country, the so-called big brown bat. It was just inside Grand Canyon off to the left of Floyd's Tomb. These are rather large bats—the only ones in this vicinity that could easily puncture a person's skin when they bite. All the other bats were deeper in the cave—no, I take it back—there was also a little brown bat near the entrance of the commercial portion of the cave. All through the rest of the cave, we saw no bats until we got to Floyd's Passage. Scattered about we have counted upwards of 45 pipistrellas.

REPORTER: Just what sort of an animal is he?

BARR: Well, it is small—about the smallest species of our eastern bats. They are rather solitary individuals, never getting together in those big hibernating colonies like most bats do. They are rather like the big brown bat in this respect. But they prefer a little bit warmer temperature.

REPORTER: They go way into the cave?

BARR: Yes. They probably go deeper in the caves than almost any other species. However, the Lost Passage is much too deep even for pipistrellas, I feel sure. They are, you might say, biological indicators of some sort of an opening, at least big enough for a bat. Well, that is the breakdown on the principal forms of cave life.

REPORTER: You have not run into anything that looks like a new species of animal or anything?

BARR: Well, it would be difficult to say. It requires a lot of research in the literature, and of course I have no microscope back here. Only on certain groups where you are very well acquainted with the taxonomy, can you make any statement of that nature. I have a few additional notes, if you want them.

REPORTER: I sure do.

BARR: As far as working on the geology—the various layers in these limestone formations vary considerably in solubility and create a number of interesting solutional features—dome pits, large passages—most of these are not generally found in small caves and are difficult to explain. We are working on several theories which might explain these various features. Keeping all of these theories in mind, we will test some of them in Crystal Cave.

REPORTER: Are you finding any unusual formations?

BARR: The gypsum crystals back at one end of Floyd's Passage are one of the two most beautiful collections I have ever seen anywhere —that goes for 250 caves or more.

REPORTER: What sort of formations are they?

BARR: They are a fibrous type of gypsum known as selenite. They appear as though they are extruded from the walls. They look like tiger lilies, for example. They come out and curve upward. These are white and tinged with a delicate brown on the outer edge, and occur at the west end of Floyd's Passage.

REPORTER: They are shaped more like tiger lilies—more than any other flower you can think of?

BARR: Yes, of course there are all sorts of fantastic shapes. Some of them are 10 to 12 inches long, which is quite large for gypsum flowers. They are not widely distributed in the cave.

REPORTER: Thank you very much, Tom.

The men with bottles were adequately carrying out their assigned mission within the time allotted. At the end of the week the expedition would have a mountain of scientific data adding up to a detailed description of one particular cave among thousands of caves.

White crayfish, common inhabitants of the underground, are found in a remote pool at the bottom of X Pit. *Photo by Thomas C. Barr, Jr.*

A bat found by Barr measures eight inches from wing tip to wing tip. *Photo by Robert Halmi.*

14

ENIGMA OF MUD AVENUE

TWENTY-FOUR HOUR EXPLORATION ACTIVITY created problems Joe and the other leaders had not been able to foresee at the start of the expedition. As the hours wore on they became aware that members of outgoing exploration parties often could not compare notes with those who had come back because of dissimilar sleeping schedules. As a result, no one was able to get a complete picture of progress, but findings were filtering in to indicate that we had a first-class mystery on our hands.

By Tuesday evening we had enough information to show that the B Trail-Bogardus area was yielding diminishing returns for the effort expended. Joe decided we would now shift our main emphasis toward solving the enigma that lay under our feet—Mud Avenue. This much we knew about Mud Avenue: it was the lowest known major passage in the system and had been used in part as a connecting link between Bottomless Pit and the Lost Passage by earlier explorers. It fully lived up to its name. A gooey layer of brown mud covered every wall and rock in the passage. Anyone venturing through it could easily be identified later by his mud-stained coveralls.

As Joe sat outlining the plan of attack on the new objective, he noticed that a few people were looking down the Lost Passage at a lone figure walking toward them. Skeets Miller's jaw dropped in amazement. He stopped talking in the middle of a sentence of his broadcast to the outer world. He hurriedly told the radio people on the other end of the line that he was starting all over again. The figure was Bill Austin, dressed in a clean double-breasted business suit and carrying a platter on which reposed a complete baked ham! How in Heaven's name had he gotten here? People converged on Bill, their eyes wide in astonishment. Skeets detained him while he resumed broadcasting:

This is Number 2 for WSM, Nashville.

There is another surprise for this group at Camp One, 250 feet underground in Floyd's Lost Passage. William T. Austin, Manager of Floyd Collins Crystal Cave, has just arrived here impeccably dressed, in gray tie, gray suit, pressed trousers, and as he walked in, he brought us a ham, beautifully decorated with pineapple, cherries and candy hearts. It will be the first real meat we have had down here because until now we have had nothing but bologna. Now the whole staff down here will be surprised, since I am in a remote section and Bill had to walk directly past me to get to them. Bill, come over here and tell me what this is all about.

AUSTIN: They say a little nonsense is good for the best of men.

MILLER: Bless your heart. This is going to please everybody down here. I wish I had a microphone so everybody could hear the shout when you walk down there about a hundred feet. Bob Halmi of *True* magazine just came running up. He is going to take a picture before Bill is lifted off the ground. Now they are all beginning to arrive. But here is Jim Dyer, who was formerly manager of Floyd Collins Crystal Cave. He was succeeded by Austin and probably knew all about this. By the way, when we first came down here last Sunday we found, set up behind what is our cafeteria, one very nice sign that said "McClure's Slop-Shop." Roger McClure, one of the advance party, got lonesome while waiting for us, so he made it. But since the women have arrived down below, the sign has been changed to "Ida and Nancy's Place." This is for Ida Sawtelle and Nancy Rogers. Now, beside the sign, some wag down here has put up a silhouette of a beautiful girl in a Bikini bathing suit. But none of the girls did this or approve it. Also coming down with the support group were two marines, who had brought blue jeans and had carefully inscribed on them their insignia, and they looked as if they were walking in their uniforms as they came in the Crawlway to bring our supplies. My only question now is how Austin came down in his impeccable gray suit? Aw-w-w, he has played a trick on us. He has come down and then changed his clothes after he got in our section of the cave. We will try to find where he has his cave clothes cached. This is Burke Miller returning you to the studio.

Austin admitted that he had brought his clothes down in a plastic sack and had changed in Floyd's Kitchen. The ham came down in another sack carried by Jim Neidner.

Austin had been considering just such a morale raiser for a number of weeks. One scheme was to bring an auto tire into the Lost Passage and roll it over the sand floor to make a neat set of tracks which would converge into a tiny side lead. The size of the Keyhole made that stunt impossible.

Low, twisting passages lead off of Mud Avenue. *Photo by Robert Halmi.*

Joe went back to the exploring business by dispatching a three-man party upstream in Mud Avenue. He hoped to determine the source of the stream in Mud Avenue and later find its outlet. Lehrberger, Smith, and Wilt climbed down into the hole in the floor at B Point and began working west. Their path lay along a slippery mud bank where footholds were difficult. The stream at this point was about one foot wide, flowing rapidly over rocks and ledges on its downward course. Twenty feet over them the jagged ceiling had been eroded into deep channels by the stream in ages past. Since their mission was to explore Mud Avenue itself, they did not turn off at the marked side passages leading to other parts of the cave.

The floor was gently rising to meet the ceiling about three hundred feet from where they had entered. Here they were again confronted with a choice. The passage split into two levels with the stream passage directly below the upper passage. Lehrberger ducked into the lower one while the others climbed up into a low tube. They could hear each other speaking as they went so they knew the two paralleled each other. For an hour they moved this way, stopping only to describe entrances to likely-looking side passages which would need exploring.

Lehrberger was yelling. He had run out of man-sized passages at a point where the stream gushed through a crevice. The others stationed themselves over the sound of his voice to wait for Lehrberger to retrace his path to a connection he had found. They heard the water gurgling below them and were impatient to start moving.

The passage pitched downward, forcing them to their bellies. They squirmed toward an opening ahead. The first man squeezed through it into a black void which proved on inspection to be a large dome, about fifty feet high and twenty feet in diameter. Crevices led out of this at several points of the compass. Lehrberger suggested they each follow one of them for five minutes then return to compare notes on their findings.

At the end of the time limit, three bewildered persons stood in the center of the room wildly flailing the air with their hands and describing where their jaunts had led them. Lehrberger had found a dome pit with a waterfall streaming down into it. His flashlight beam was unable to penetrate the darkness above, so he had no idea at all of its height. Smith reported finding three small dome pits about twenty feet high, each with water falling into them. The water disappeared between breakdown rocks on the floor. Wilt announced he had found a similar series of pits containing falling water. They agreed that this was the source of the Mud Avenue stream, and returned to camp with news of their discoveries.

With one half of the engima solved, Joe sent Thierry and Charlton to examine one particularly promising side lead to the left in Mud Avenue, an area untouched by any previous exploration. They climbed down into Mud Avenue and quickly came to their assigned passage. Thierry boosted Charlton up five feet into it, then grasped Charlton's hand for an assist up. At the end of thirty-five feet of easy crawling, they were amazed to look through a hole in the ceiling and see a room of enormous proportions. It was as big as the Lost Passage, they thought. Their hearts pounded as they scrambled up a pile of boulders to the lip of the hole. On climbing out they found that their discovery was in no way smaller than the Lost Passage. It *was* the Lost Passage, and the glow of Camp One around the bend made them feel a little ridiculous.

But just ahead of them they noticed a tiny hole in the wall at the floor level. It led in the direction they had been going, so they squeezed in, finding it not much bigger in diameter than a small wastepaper basket. After ten minutes of tight crawling they reached a small pit at a distance they estimated to be about one hundred feet from the Lost Passage. The pit dropped twenty feet to its floor, but they found no way to chimney down into it.

Just as they were about to give up, Charlton found a squeeze-lead going off from the right hand wall of the pit about four feet away from them. A search for footholds to gain access yielded nothing; however, Charlton did find a rock projection jutting into the pit near the ceiling. He circled his arms around it and pulled his body out into the pit. At the last minute he kicked off with his foot and swung his body in a downward arc, ending with a toehold in the new passage. After struggling for a few moments he worked his way to safety.

When Thierry reached the other side he found Roy squeezing his way into a tight vertical slit-like passage. Thierry doubted his ability to force his large bulk into the slit to follow the short, agile man ahead of him, but he made the attempt and succeeded by exhaling and pushing with all his strength. He caught up to Roy at the brink of another pit, smaller than the last one. They easily stepped over this and continued to push and pull their way on for several hundred feet. They began to see formations as the passage widened out. Stalactites and ribbon deposits hung from the ceiling. Some were colored a delicate tan at their bases and tapered down to pure white on the ends. Drops of water hung poised on their lowermost points.

Their tour ended when they climbed down into the spring basin containing the tube passage leading to Floyd's Kitchen. In a few minutes they were back in camp to report their findings. We could chalk off one more passage in the labyrinth.

On Wednesday, Austin and Neidner volunteered to help in the exploration of Mud Avenue. Lehrberger suggested they go downstream to see if they could find the destination of the water. Perhaps they could find the large underground river which emptied into the Green River at Pike Spring outside.

The course of this river had been traced many years ago from a point in Colossal Cave, to the south, to Pike Spring. When the Louisville and

The Shillelagh hanging from the ceiling in Mud Avenue delights photographers. This rare dripstone formation is covered with a helictite fringe, contrasting with its dismal muddy surroundings. *Photo by Robert Handley.*

Nashville Railroad operated that cave, they had dumped a quantity of dye into the river. A few hours later it stained the water emerging from the mouth of the spring two miles away. If the two locations were connected by a straight line, it would pass within a thousand feet of known passages in Crystal. We knew where Mud Avenue water came from, but where did it go?

Austin, Lehrberger and Neidner pushed downstream, following an increasingly muddy passage. They climbed over mud-caked boulders, liberally plastering themselves with "goo." The stream ran in a V-shaped hollow between two high mud banks, offering them no choice except to walk in the water. They sank in up to their ankles as they steadied themselves on the slippery walls. They looked momentarily at a higher-level passage, the same kind of junction Lehrberger had encountered near the beginning of Mud Avenue, and decided to follow the lower one.

Bill Austin stopped the party. He reached toward a ledge where a burned-out flashbulb had lodged. No one before had ventured down Mud Avenue, and none of the explorers prior to Austin's time had taken pictures in the lower levels. Austin told the group that the bulb had apparently been washed there after either he or Dyer had dropped it back beyond B Point on an early trip. It would take a large quantity of water to lodge it so high on the wall; he remembered the flood which had raised the level of Green River during the winter of 1952-53. He asked the others how far they thought the flashbulb was below the level of the Lost Passage. Their consensus was about twenty feet. He made a note to do some checking later.

They were duckwalking in the icy water now and the ceiling pinched down on their heads. Ahead they heard a faint sound, like that of water plunging over a cliff. They could go forward only by squirming along flat in the water with their heads barely clearing the ceiling. The idea didn't appeal to them, but they compromised and decided they would press on only as far as possible on hands and knees in the hope that the passage would not continue to grow smaller. Austin called back to the others. He was stopped by a mud wall punctured by a small sewer tube into which the water plunged, not big enough for a man to enter.

They retreated to the high passage and again went forward. On the floor of this squirmway was a four-inch layer of mud, the consistency of axle grease. It oozed into their clothing at their wrists and pockets. They could feel it forcing its way into their shoes, but the sound of falling water was louder now. It drew them like a magnet as they picked up speed in spite of their unpleasant surroundings. A large rock blocked their way. Air streamed past, cooling their hot faces. Austin pushed on it, watching it move, then settle back in position. Two of them rocked it back and forth in its mud socket, watched while it teetered for a split second and disappeared from view. They crawled after it into a high pit with a waterfall issuing from the top. Above them, a much larger passage eight feet wide and twelve feet high continued in the direction they had been heading, but a sheer mud wall denied access to it. Austin pricked up his ears. Over the waterfall din he could hear a deeper noise, a faint rumble of water. The others heard it too. The underground river? Their flashlights shone into the dark mouth of the top passage. Lehrberger estimated the distance up to it at twelve feet.

They rested to calm their excitement. They had been gone for three hours and now their teeth were chattering. Even if they had brought ropes, they were in no mood to use them. When Lehrberger proposed a hasty retreat, no one argued with him.

They arrived in camp dripping with water just as Skeets Miller had made contact for his scheduled broadcast:

> This is Number 1 for WSM, Nashville:
> Explorations are still continuing here beyond Floyd's Lost Passage and almost hourly some group comes back reporting some find of minor or possibly major importance. Just as I am talking to you, Bill Austin, second in command of the Expedition, has come back in wet clothing and Jack Lehrberger is with him. They have calmly

announced they went swimming down here underground. What is this, John, about swimming? Where did you go?

LEHRBERGER: We followed Mud Avenue downstream and we finally got to a point where we would have to swim to go any farther so we went to a little upper-level passage and continued on, but we were stopped by a 50-foot pit and dome affair.

MILLER: Did you actually get wet or did you fall in the water? (Snickers in the background) I ask if they got wet while Bob Handley is pointing to their clothes caked with mud. Thanks very much, and get into some dry clothes. This goes on constantly down here under the ground on this our 4th day of sending programs to you from Floyd Collins' Lost Passage in Crystal Cave, Kentucky. This is Burke Miller speaking for WSM, Nashville.

When Miller had finished, he looked over the haggard trio drinking hot coffee. It was difficult to tell the color of their clothing because it was hidden behind an almost unbroken façade of mud. Their cheeks and foreheads were smeared with it.

Bill Austin checks with his wife, Jacque, on Green River flood data after his trip down Mud Avenue. *Photo by Robert Halmi.*

Austin revived somewhat and contacted his wife on the telephone. She dictated some figures to him as he asked for them, and he marked them down on a mud-smudged scrap of paper on which he did some rapid calculating.

"I think I've got it," he said. "According to my figuring, the Lost Passage is about 490 feet above sea level. We know from the topographic map that Base Camp is about 750 feet above sea level. That puts us 260 feet underground." Skeets was curious to know how he had arrived at that conclusion. Austin explained finding the flashbulb on the ledge. It must have been floated there by high water backed up from Green River during the flood, he said. Through his wife, he had obtained information from his files establishing that at flood stage the Green River had risen to 473.25 feet above sea level. This he assumed to be the elevation of the flashbulb. The party's estimate that the Lost Passage lay almost twenty feet above that point gave a rough elevation figure of 490 feet for Camp One.

As Expedition Leader, Joe felt that he could not, with a clear conscience, send another party back to find the underground river. One look at the last party out showed him the unmistakable marks of fatigue on their faces. The discomforts were obviously too great to endure so long as we had plenty of unexplored dry passage to search out. The enigma of Mud Avenue would be his gift to future speleologists.

15

SMOKESCREEN

I OPENED MY EYES and sat bolt upright in my sleeping bag. There was no light from the direction of Camp One and no noise except the dripping of water from some place to my left. Could I have slept like Rip Van Winkle while everyone else had left the cave? Then I remembered that the bend in the passage would block off any light from camp, so I breathed deeply to shake off my sleepiness, then switched on my flashlight to look at my watch. Six o'clock; but was it morning or evening? When had I gone to sleep? I consulted the tag at the end of the sleeping bag. R. BRUCKER IN SACK 10:30 A.M. TUESDAY, 16 FEB. How long would a person sleep after going continually for fifty-three hours? It must be morning.

I washed my hands and face in the spring and walked into camp for breakfast. My breakfast was everyone else's supper, for it was actually six in the evening on Tuesday and still the third day of the expedition. Earl Thierry and Roy Charlton were there eating, and they looked unearthly in their beards and mud-crusted clothes. The log filled me in on some of the events of the last few hours.

At eleven o'clock this morning, shortly after I had gone to bed, Lawrence had dispatched a support party to take empty Gurnee cans to the surface. Ida was leader, with Ken Perry and John Spence completing the detail. Ida protested that she didn't know the way, but Joe said it was easy, all they had to do was follow the phone lines. What Joe didn't know, however, was that Ida had originally come in by way of a devious route from the bottom of the Bottomless Pit to Mud Avenue and from there into the Lost Passage. Since this was the only route she thought she knew, she led the party over as much of it as she could remember. They made the wrong turn off Mud Avenue and wound up an hour later at Floyd's Jump-Off. Not recognizing this, they retreated again to Mud Avenue and continued along it to another side passage.

This time Ida recognized the way and soon had the party standing in the bottom of the Bottomless Pit. A support team was passing overhead, so Ida yelled up to them asking where the phone wires were. "At the top," assured Luther Miller. By the time Ida had led her group up the tricky climb, the support party had vanished in the night. They yelled but no one heard. Resolutely, they followed Joe's instructions from here on.

Within an hour the astonished party walked back into Camp One . . . still following the phone wires! Ida summed up the sentiment of the group when she said, "Now we know the phone wires run both ways!"

Also during the afternoon an exploring party had come back from B Trail, another had gone to X Pit, and a third up Mud Avenue. A supply team had come in and a survey party had gone out. Ken Perry and John Spence had set off again leaving Ida behind. She went with a party led by Bill Austin to attempt to get to a passage leading off from the far side of Frenchman's Pit.

"What do you think we ought to do with all this trash?" asked Earl, looking at the pile of discarded paper towels, cereal boxes, and old newspapers.

Nancy Rogers looked at them a moment, then said, "Why don't you take them on down the passage and burn them? The air seems to be blowing toward the north end. Smoke should go right out."

"I'll take them down beyond the breakdown. That ought to be far enough." Earl scooped up the trash in his arms and set out.

In ten minutes he was back. "Got a real fine blaze going. It's a shame we don't have any marshmallows." But Nancy wasn't listening. She was sniffing the air. The faint odor of smoke drifted into Camp One accompanied by wisps of smoke driven by a shifting wind. It was becoming difficult to see through the increasing haze, and someone was coughing down the passage. My eyes were watering now as I watched a halo form around the gasoline lantern. Charlton, on the other side of the room faded from view, then he coughed and walked over to us.

"Think it's about time for a little survey trip to start out, don't you, Earl?" said Roy. We groped for our compass and chain in the pile of equipment and beat a hasty retreat toward B Point two hundred feet south of Camp One.

Here, where the air was somewhat more breathable, we started our standard survey procedure. How unusual it was to measure off hundred-

and two-hundred-foot shots, a far cry from the twisting B Trail and Bogardus Waterfall Trail. A pleasant surprise to us. In three quarters of an hour we surveyed to the crack in the ceiling where Roy established station K, choosing this designation because we were in Floyd's Kitchen. Earl and I scrambled up the muddy tube behind the spring and soon stood over the crack where we could see Roy's light below. Earl uncoiled the chain again and lowered the end through the crack. In a moment Roy cried, "Chain!" from down below. Earl stretched the chain taut, checking it with the plumb bob to see that it was as near vertical as he could make it. Station K-1 was marked in carbide eighteen feet directly over station K.

Roy Charlton joined us as we surveyed along the supply route up through a small hole in the top of a breakdown. We inched along over a series of thin flat slabs, about as large as ping-pong tables. Then we stopped.

"What's this passage?" I asked. I had crossed it several times on the way to the Lost Passage, but really had never studied it closely.

Roy was quick to answer, "This is kind of a flat room. It goes down that way a few hundred feet and ends in a rather large room. Not many people have been in it."

"Let's survey it then," said Earl. "This is one of those places everyone goes through, but nobody ever explores. Too obvious. Most of these cave explorers want to move rocks and crawl for miles."

Remembering Roy's remark, we dubbed the passage the "Flat Room" and started surveying from one F station to the next. We had no idea where the survey or the passage would take us, but it looked exciting. The passage was very large; at our first station the ceiling was ten feet high, and the passage about fifty feet wide. The floor was covered with a peculiar crusty red sand mingled with sparkling gypsum. As we moved along, the sand fill began to close the gap between the ceiling and floor, but at the same time the passage was getting wider. Seventy feet wide at F-2, it continued that width beyond F-3, located more than three hundred feet farther down the passage. At that distance Roy's light bobbed like a mere pinprick of light in the darkness. So this is the Flat Room! The excitement increased as we pressed on, each of us amazed at the avenue's broad vista. It was twice as wide as the Lost Passage in some places, and it was eighteen feet above it. Perhaps we were on the trail of the legendary room in Crystal which earlier explorers had described as dwarfing the Lost Passage. We raced onward, following the footprints of three people, Roy's and two others.

We met Roy in a large open place in the passage. The sand floor had mysteriously stopped, and as we looked down into the broad basin we saw large chunks of breakdown littering the room. We moved down to join Roy. The ceiling was now about twenty-five feet over our heads, and the walls enclosed a room about eighty feet in diameter. At places, particularly inviting places, passages apparently led off into more cave. Roy was sniffing the air. "Smell that?"

"Yes," I said, "it smells like something burning. Where's it coming from?"

"Must be from the Lost Passage—my little bonfire," admitted Earl.

"So you're the one responsible for the smoke hanging over Camp One!" said Roy.

"But we lost the scent of it after we left B Point, and we haven't smelled the smoke since," I said. We had a first-class mystery on our hands. Hurriedly we scrambled about the room, sniffing in every opening, even turning over loose rocks, and furtively turning to see if any of the others had found clues to the smoke's source.

"Hey!" yelled Roy. "Over here. It's pretty strong in this side passage." We converged on him and stopped to sniff the pungent air. "I'll check out this lead. Give me about ten minutes." He disappeared into a hole under the ledge. Earl and I finished describing the room in writing for the survey notes, then systematically worked our way around its perimeter. At the opposite side from which we had entered, the sand floor again resumed, sloping gently up to within a few inches of the ceiling. It looked as though one could go on if he weren't averse to digging beyond this barrier of sand where the passage opened up again and appeared to continue. To the left we found a small crawlway following the wall. This we would save for Roy's return. A search behind a large breakdown block revealed nothing. At a point just over the passage where Roy had entered we discovered a splendid display of gypsum flowers. Some curved like delicate orchids; others pointed downward with featherlike spines. And then we saw something spectacular and rare: a white, featherlike gypsum flower covered a stalactite hanging from the ceiling. The stalactite is a characteristic wet-cave formation. As the cave-development cycle progressed, the stalactite had become dead, or dry. Then these dry-cave formations had taken over and somehow developed on top of the stalactite. Behind this wealth of formations a passage led upward at a steep angle.

Roy was back now, reporting that the passage was about two feet high, twelve feet wide, and appeared to double back under the Flat Room. We continued the Flat Room survey into this passage for about eighty feet, to a point where breakdown blocked it. Somehow, the smell of smoke was not present at the back of this passage, so we were unable to solve the mystery to our own satisfaction immediately. When the map had been finally plotted we were able to see that this passage would have joined a side lead off the Lost Passage had we been able to push on sixty feet farther.

Roy struggled into the tiny passage at the end of the room and shouted back that it ended in a small room only three feet high, three feet wide, and seven feet long. He reported that he could see about twenty feet under a ledge to a point where the passage became larger, but without digging tools further progress was impossible.

We retraced our steps along the Flat Room, this time concentrating our attention on the south wall. We had noticed that in some spots the passage became wider, the ceiling dipping from sight behind low sand dunes. Behind one of these we discovered a passage, which proved to be a cut-around, connecting into the Flat Room again about a hundred feet farther east. Earl Thierry crawled off to explore another and was gone for fifteen minutes before he returned. He came back announcing he had been traveling in virgin passage! We scrambled after him on the cool sand, after sighting and measuring from the nearest established F station. We finally surveyed our way into an area apparently below the level we had been in, for the red-sand floor had gone, and was replaced by a rock floor. In a few more moments we came upon an arrow.

This was a traveled route, all right, but we had no idea in the world where we were. We made one sight into a side passage and I crawled back to explore it, discovering a chain-like series of pits and domes. Voices were murmuring above and behind me.

"Hello," I yelled.

"Hello," came back distantly.

"Where are you?"

"Up here . . . where are you?"

"Down here," I called.

"Where is 'down here'?" came the voice.

"I don't know."

"We don't know where we are either."

"Are you lost?" I asked.

"No . . . we *know* how to get back to Camp One from here, but we *don't* know the name of this place."

"We're in the same boat!" I yelled. That was the trouble with exploring Crystal Cave. You never knew exactly where you were at any given time, but you were almost never lost. Communication became a problem in a situation like this, for even when recounting experiences, the lack of common names or common sharing in the experience made complete understanding impossible. One party is on its way from somewhere to Camp One. Another party, ours, is on its way from Camp One to somewhere. No one really knows where those somewheres are, and they won't ever know unless the entire cave is mapped. I rejoined Roy and Earl.

Earl asked, "What was all that yelling about?"

"Just some phantoms passing in the night."

A hundred feet farther on we came to a place where the ceiling opened up to some upper levels. "This looks like the Big Breakdown," said Roy. He turned up his carbide flame and scanned a huge boulder lodged at a crazy angle in the shaft above us. "Yes, this is it, this used to be the main route into the cave; I know where we are now." It was as if the sun had come from behind an opaque cloud, for we were on familiar ground. Roy suggested, "Let's survey up the Big Breakdown to the supply route, and from there it will take us about thirty minutes to link up with Station F-1. That'll give us a good closure."

Closure is the magic word in cave surveying, for without it there is no way to check the accuracy of a survey. Once you have surveyed around a large loop—and this one was about a thousand feet—you can plot all the points and find out how nearly the plot closes. The distance between the beginning and the end of the loop when plotted on a map is the error of closure. After the morning's surveying activities had been plotted, we found an error of four feet, a tolerable error in cave surveying, but not so good by surface surveying standards. But the purpose of a cave survey is to arrive at an understanding of the cave passage lay-out and not to subdivide it into real estate. We headed back to lunch, or supper, whatever it was. At any rate, we ate the chile and oatmeal at one o'clock Wednesday morning. Earl's smoke screen was now only a memory.

Shortly after eating, Lou Lutz joined us to survey from station B to the north end of the Lost Passage. Since we needed another man to help with the chain on the long shots we would be taking, Lou was a welcome addition.

In the Lost Passage we could often see as far as three or four hundred feet before a bend blocked our view. For this reason we swept down the passage at a rapid rate, and my note taking was the chief cause of delay, for I drew in passage widths, places where side leads departed from the main route, and plotted the locations of prominent breakdown blocks. About four in the morning we were at the bottom of a large pile of breakdown that appeared to fill the passage. Roy scouted over it and yelled for us to continue the survey up the pile. We entered a huge room, second in size only to Grand Canyon on the upper levels. Its long axis lay at right angles to the direction of the Lost Passage, and we measured it to be a hundred and fifty feet long and ninety feet wide. The ceiling we estimated to be about forty feet above the breakdown floor. It looked as if this was the intersection of two large cave passages, so we hastened to explore for leads.

Roy led the way toward a precipitous ledge about half-way up on the wall. We could see that his chosen path would lead us to a passage opening at the western extremity of the room. Before climbing on the ledge we stopped to assess the hazards of making the ugly-looking traverse. The ledge was a sort of shelf projecting outward from a slanting rock wall. About a foot wide, it was covered with mud-coated scree, imbedded in a gray sticky clay-like substance. From the point where the wall met the ledge, it tilted at a slight angle toward the center of the room. Over the outer edge it was a twenty-foot drop to the jagged boulders below.

My rubber cleats bit into the soft mud as I followed the others. Because of the slope of the ledge, we were careful not to lean toward the wall, a natural tendency that is counteracted only by constant awareness. Leaning toward the wall would shift body weight toward the high side of the angle, imposing a sideways thrust on feet, and sending the explorer sliding over the brink. By remaining as upright as possible, our centers of gravity were directly over our feet, aiding them to sink in to hold us firmly. Each of us stopped momentarily to eye the rocks below, not in terror, but because this is a standard cave-exploring practice. When attempting anything the least bit risky, we formulated a "panic plan" of emergency procedures we would follow. Every explorer works out his own, and that is what we were doing at each stop.

In case of a slip, I resolved to avoid a particularly jagged pinnacle, and instead, to aim for a relatively soft flat area between two of the larger rocks. We moved again, testing each foothold before we would trust our weight to it. Charlton reached the half-way point and yelled back to us that the going was becoming easier. Nobody answered him, because the silent file of men were too busy looking after their own safety to have their attention diverted for even a moment. Ahead of me in the glow of my light I saw that the quantity of mud had diminished and that we were walking on fairly dry rocks. Soon the ledge widened to three feet.

We edged our way across a canyon in which a muddy trickle of water ran, and finally came to a dead end. Red sand, the same kind we had encountered in the Flat Room, sloped gently up to the ceiling, barring further progress at a place we had not been able to see before. Earl retraced his steps and chimneyed down into the canyon.

"You're going to get wet!" yelled Roy.

"I've been wet before, but this looks like a good lead," said Earl. "If it doesn't go any place, no one will have to do this again." He followed the canyon out of sight, and the rest of us lighted cigarettes, shined our flashlights into the great vault, and speculated on the enormous solution process that had carved out this spacious room several hundred feet underground. The only other time I had experienced a similar feeling was when I stood inside the perisphere at the New York World's Fair in 1940 and gazed at the free space enclosed by those curving plaster walls.

Earl was back, covered with mud which he attempted to scrape off his coveralls with a thin piece of rock. "It's a stoop passage back about fifty feet, and then it pinches down to a muddy crawlway. I didn't see any sense in following it as long as we have all this fine walking cave to survey and explore." We accepted his decision and headed for the other side of the room. Back across the ledge we went, repeating the same cautious process: take a step, test it, look around, look down, shift weight, look, take another step, and so on. Everyone breathed easier on the other side. We picked our way across the breakdown and arrived at a smooth limestone wall. There appeared to be no chance of going any farther, so we gingerly worked our way along the wall.

"Hey," yelled Roy, "here's a way down!" He was shining his flashlight into a black void between two pieces of breakdown and the wall. "Let's go!" Earl kicked the rocks to see if they would stay put; they didn't budge. He

There are many pits and domes in an area near the Flat Room. Passages leading out of these pits at different levels are a challenge to both the explorer and the rock climber. *Photo by R. J. Richter.*

placed his hands on the rocks to lower himself. We could see his light bobbing about as he slowly edged his way down to the bottom where he called for us to follow.

We were standing in a large passage, of the same general character as the Lost Passage and on the same level. We tried to reconstruct the chain of geologic events as they had probably happened. At one time an extensive water channel had run east and west forming the large room. Simultaneously, the Lost Passage had been forming on a north and south line. It appeared as if the ceiling of the Lost Passage directly under the room had collapsed, and the ends of the large passage above had been filled by red sand deposited later. We surveyed for a hundred and ten feet in the north direction, finally coming to a magnificent flowstone-covered breakdown. Stalactites hung from the ceiling, and a veritable waterfall of cream-colored flowstone ran down in smooth mounds to the floor. It was one of the finest displays of its kind in the cave, and would rival those found in almost any commercial cave.

As we looked at it, a strong breeze sprang up from the direction of the flowstone. Roy needed no encouragement. He ran up the wet flowstone slope and disappeared behind a large rock. We followed him, but he was already out of sight.

"I don't think he's going to get very far. This is probably the end of the cave," said Earl. "I'll go in with him and survey our way back out." He

Dr. Wanger's Camp One dispensary does not have the white, immaculate appearance common to medical installations, but it is effective in easing headaches and insomnia. *Photo by John Spence.*

took the compass and chain and left. We could hear him yelling to Roy who, we learned, was flat on his stomach in a wet crawlway and rapidly running out of cave.

"The breeze is strong back there," said Roy on returning. "Again, we'd need digging tools to get anywhere. I'm going to mark that one for the next time I come down into this cave." He had discovered a clear lake about thirty feet across and a small crawlway leading off the opposite shore. The passage led to a series of small rooms. He said he could see into another room, apparently larger, beyond the point at which he stopped. One day this may lead to a new entrance to Crystal Cave if persevering cave explorers keep working.

We rested for a few minutes and then walked leisurely along the half-mile passage to camp.

About ten o'clock in the morning Roger McClure appeared, rubbing sleep from his eyes and wondering what was up. We told him of our last few hours' activities. Roy and Earl decided they would explore up Mud Avenue, and I gathered together a party consisting of Lou Lutz, Roger McClure, and Phil Smith. Our aim was to extend the survey from the top of the Big Breakdown as far back along the supply route as we could go before lunch time, a time I had set as an arbitrary bedtime, sleeping bag or no sleeping bag. The surveying covered old ground that we knew thoroughly, and proceeded without incident worthy of mention. We stopped within several hundred feet of the Bottomless Pit before we headed back for lunch.

I found myself speaking incoherently after lunch, although my thoughts were quite clear. Doc Wanger came and looked me over with a curious, professional eye. He put on his best cave-side manner and tried to draw me out concerning how I felt, how long I had been without sleep, and how regularly I had eaten. I told him I was on my way to bed at the moment and felt rather tired. He suggested I take one of his sleeping tablets, an intriguing idea; though I felt no need for it I took it out of curiosity. Even before I had crawled into my sleeping bag and written my name on a call tag at my feet, my eyelids were closing. It was noon Wednesday, the fourth day of the expedition when I dropped off to sleep.

16

BRAND-NEW CAVE

A T NOON WEDNESDAY other things were taking place while a dozen other cavers and I slumbered underground. Joe Lawrence consulted with Doc Wanger on the condition of Marguerite Klein with a view to evacuating her from the cave. She had been in her sleeping bag ever since her arrival with the "Fourteen-hour Nightmare" party two days ago. Marguerite had been feeling well all morning and it was the Doctor's opinion that she would be able to stand the trip out with little difficulty. Joe scheduled departure for three in the afternoon, and he decided to go along as party leader, since several matters needed his attention at Base Camp.

Dr. Feder, Ackie Loyd and Joe departed with Marguerite for a leisurely trip to Base Camp. They used nylon safety ropes to secure her across the Bottomless Pit and some of the more treacherous canyon traverses.

When their party reached the Keyhole, they found an oven and some forlorn cans of biscuits abandoned by a support party. The oven, a small sheet-metal box twelve inches high, was intended for use with a gasoline stove for baking the biscuits. Joe crawled through the Keyhole and found out why it had been abandoned. The twelve-inch oven was too large to pass through the Keyhole! A mason's hammer lay near the oven; evidently someone had brought it to enlarge the Keyhole by brute force. Joe swung the hammer at the rock floor of the Keyhole. A chip flew—a very small chip. A few more blows confirmed that he could never hope to cut away enough of the hard rock to allow the oven to pass through.

If he couldn't make the Keyhole larger, he could certainly make the oven smaller. He swung the hammer at the oven, banging it again and again, driving the top down in a crumpled mass. Finally he pushed the eleven-inch oven into the Keyhole until it wedged tightly in the opening. Using the hammer, he tried to drive the oven through to Ned Feder on the other side. His energetic efforts only wedged it more tightly into the lozenge-shaped crevice. It wasn't going through, he thought as he tried to pull the oven back to him. The oven was stuck. It looked as though Ned Feder would have to open the oven door and crawl through if he wanted to get out of the cave.

Ned kicked the oven; it moved a little. He kicked it again and it landed in Joe's lap.

Joe used the hammer again to make the oven still smaller. After much noisy hammering, he placed the ten-inch oven in the Keyhole, finding it a snug fit. Again he tried to drive it through, and as he pounded, it slowly squeezed into the Keyhole. At last the oven, hardly recognizable by this time, popped through into the passage beyond.

They left it for a support party to pick up and they proceeded toward the entrance. Fatigue began to overtake Joe. He assured the others that Scotchman's Trap would be around the next turn, but around it was only more Crawlway. The Trap couldn't be more than fifty feet away. Joe covered

Bill Wanger attempts to enlarge the 10-inch Keyhole to accommodate a 12-inch oven. Every explorer and all equipment and supplies had to pass through this bottleneck to reach Camp One. *Photo by Charles Kacsur.*

two hundred feet and was still in the Crawlway. He longed to stand up —to get away from those walls that crushed in from either side snagging his clothes with their knobby fingers. His desire to be out of the Crawlway caused him to increase his speed, to get out of this physically uncomfortable and nerve-wearing place.

"Slow down, Joe," Dr. Feder ordered. "You're going too fast for Marguerite."

This brought Joe back to proper perspective. He traveled at Marguerite's speed the rest of the way to the Trap.

Marguerite was taken to Mammoth Cave Hotel where she could get some rest. After the expedition Marguerite expressed her feelings quite clearly on what happened to her. "Those reporters were just sitting around waiting for another Floyd Collins episode, or at least a broken leg or cracked skull. No one obliged them, and things were getting dull from their viewpoint. A real bed in the hotel felt wonderful after sleeping in the cave. I was hounded for two days by reporters and photographers taking pictures of me hanging up clothes, drying dishes, and even in a sleeping bag getting my pulse taken by an intern. They told me to cooperate, so I couldn't do anything about it. Every time I would go into a store in Cave City the proprietor would say, 'Oh, you're the one they had to carry out of the cave!' Boy, I was really fed up with those damn reporters."

After getting Marguerite off to Mammoth Cave Hotel, Joe immediately went into a huddle with Bill Austin and several of the staff officers to plan the expedition's future operations. There was a knock on the door.

"Who's there?"

"Ackie Loyd."

"Come in, Ackie."

"I wish I were back in Camp One."

"Why? What's the trouble, Ackie?"

"This information business is too much for me. There's a reporter pumping me for the story about Marguerite Klein."

"He can't do that," said Burton Faust. "No one is to give interviews without permission from me."

"I know, that's why I . . ."

"Is Ackie Loyd here?" A newspaper reporter poked his head through the doorway without knocking.

"Yes he is, but he's busy," said Austin.

The reporter came into the room. "I want to find out about the girl he brought out of the cave."

"You'll have to wait until we're through with Ackie," snapped Austin.

"I can't wait. I have a deadline to meet."

"Here's your story," Joe said, "Marguerite came out of the cave under her own power. She seems to be in good condition. She's resting at the Mammoth Cave Hotel now."

"But I want my story from Ackie."

"If you'll wait outside we can finish our business; then you can talk to Ackie."

Reluctantly, the reporter left. Ackie soon followed after him with instructions to give him any details he needed.

Joe then began to discuss communications with Bob Lutz, but Lutz was interrupted by a phone call. When he hung up he said, "You gentlemen may be interested in knowing that our reporter friend is now interviewing Dr. Wanger in Camp One by telephone."

"What's he trying to pull?" growled Faust. "He knows he's not supposed to do that."

Bob Lutz rose, pulled his wire-cutters out of his pocket and walked toward the door. "I'll fix that," he muttered as he disappeared in the night.

In five minutes Lutz was back. "We've just had a total communications failure. It's impossible to reach Camp One. All we have to do now is wait for it to be reported by the operator."

Ten minutes later the phone rang. The operator said, "Something's wrong. I can't reach Camp One!"

"That doesn't sound reasonable," Joe said. "Are you sure?"

"Yes, I can't get them on either circuit."

"Just a minute, I'll let Lutz talk to you."

Bob spoke briefly to the operator, then dashed out of the room.

Bill Austin and Joe were able to finish their planning without further interruptions; then Joe walked down to the communications tent. Bob Lutz was working busily over the switchboard while the confused operator stood by.

A few feet away the reporter was speaking excitedly to his office over the outside line, ". . . yes, I said a *complete* communications breakdown. Camp One is cut off from the outside world! I don't know what's happening down there!"

"Serves him right," said Joe as he walked out into a cold cloudy night and headed for a bed in Cave City.

Back in Camp One after Joe's party had left the Lost Passage, Wednesday afternoon and evening passed in the lower levels of Crystal Cave with the log showing six exploration parties departing for various places, and some of them returning with no discoveries of note. Two support teams swept in with full loads, then vanished into the night of Crystal Cave, and around-the-clock feeding continued in Ida and Nancy's Place, with Ida working over the hot stove.

About 9:30 P.M. Earl Thierry arose and walked into camp. Ida explained that before he could be fed he would have to make a trip to the spring for water for washing dishes. Taking two large pots he departed into the darkness as so many of us had before.

A sharp metallic clatter brought everyone to his feet, and all eyes turned in the direction of the spring. Men raced at top speed to see what catastrophe had befallen Earl. By the sound of it, he had tumbled down the ten-foot well. Speed was necessary to bring all possible aid.

Thierry sat in the pool at the bottom, blinking up at the would-be rescuers. The pots floated beside him. "What happened?" asked one of the group.

Earl looked bewildered as he spoke. "I'm not sure. I filled the pots with water and had one in each hand when I started to climb out. About halfway up my foot slipped."

"Are you hurt?"

"No, just bumped elbow and knee. I'm all right." Thierry scrambled out of the pool and up the climb that was difficult enough, with only one pot of water. Doc Wanger swung into action, plastering Band-Aids over the scrapes, and soon Thierry was ready to go again.

It was about 1:30 A.M. when I awoke with a clear head but a feeling of complete bewilderment as to whether it was morning or afternoon, or even what day it was. I learned at breakfast that it was 1:45 on the morning of the 18th of February, a Thursday. There was some talk about the phone having been out of order for several hours, but everyone thought it was probably the result of a switchboard operator falling asleep.

Jim Dyer came up to me after I had eaten and wondered if I was equal to trying a pit crossing. I asked for details.

"Well," he said, "I think Luther knows where Frenchman's is. It's the one the climbing team tried a little while ago. They didn't have enough equipment to tackle it so they came back. You go up Mud Avenue and cut over to the bottom of the Bottomless Pit. The passage going to it is somewhere near the bottom of the Big Breakdown."

"Who do you want to go along?" I asked.

"Bruck, I thought maybe you and Luther would lead. Lou Lutz is eager to do some climbing, and Doc wants to take some pictures. Bill Wanger wants to go too."

From Lou I learned that we would be going to the same place that the climbing party had tried yesterday, and that the "phantoms in the night" had actually been that party returning from the Pit.

I assembled the team and checked over our equipment. We had about a hundred and fifty feet of nylon climbing rope and Lou's rope waistband sagged under a full load of pitons and assorted mountain-climbing hardware. We signed out in the log and were on our way.

Near B Point we dropped down into Mud Avenue, slid down a mud slide, and walked along the trail paralleling a small stream. Luther eyed the passages as we went by, looking for the one that would take us to the Bottomless Pit.

"We must have passed it!" he said. "I've been through it a dozen times, but I'll be darned if I can remember which one it is." We could be of no help in this, so we simply followed him. At last he found a passage that struck a chord in his memory. Getting into it meant chimneying up about ten feet into a narrow crack, then about fifty feet sideways through the crack. Ten minutes later we stood in the bottom of the Bottomless Pit looking up. The narrow ledge we had crossed over so many times looked infinitely more terrifying from the bottom, for our combined lights enabled us to see it clearly. Crossing from above, there was room for only one person at a time, so only one light penetrated the dark depths—perhaps fortunately so.

Doc Wanger wanted to take a picture while we rested, so Luther suggested that he get a dramatic one. Luther rounded a corner, and presently appeared on a narrow ledge jutting into the pit at a distance of about 12 feet from the floor. If it were framed right, it would be almost as impressive as the photographs of Indians sitting on the rim of the Grand Canyon of Arizona which adorn travel folders. To add to the interest I climbed up the wall and grabbed Luther's foot, as though he were in the process of hauling me up to his precarious perch from one even more dangerous. Lou stepped in, imbedding his fingers in a tiny crack, hanging on for dear life. Halmi would have given his eyeteeth to be here. As we posed we all told Doc we wanted a copy to show our grandchildren when they asked what it was like to go cave exploring for a week.

"Can't you hurry, Doc?" asked Lou. "Pretty hard to hold on here."

"Just a minute . . . let me get this flashgun wire plugged in . . . oops . . . now . . . almost got it. . . ."

"What is it now?" asked Luther.

"The bulb won't go off!" explained Doc.

"Get another!" yelled Lou. We could hear the sounds of frantic scuffling as the Doc searched through his gadget bag for another bulb.

"Got one . . . there . . . ready now?"

". . . for an hour. Shoot! . . ."

". . . . . . damn!"

"What now?"

"This one won't go off either. I think it's a short in the wire. If you can wait a few minutes I think I can fix . . . DAMN . . . right in my face!" A flash had filled the pit for a brief instant. Doc had crossed the wires somehow and fired the flashbulb prematurely. "Can you guys hold that pose for a few more minutes? I'm going to try it with the magnesium ribbon."

My arm was aching from holding on. Lou was quivering with the tension he had been exerting. Luther's leg was getting a cramp in it. A few more moments and the Doc would get not only a spectacular picture of three people falling through space, but also a batch of patients. We were about to give up when we heard the Doc's cheery voice: "I can't get the ribbon to light. Maybe we can get the shot on the way back." We didn't think we'd come back this way very soon.

We squeezed through a crack in the wall, then on our sides through a narrow crevice. We were at the bottom of the Big Breakdown. I felt proud at recognizing the place and the evidence that we had been there before in the form of brand-new survey stations.

"I think Frenchman's Pit is down this passage somewhere," said Lou. "Seems to me we've been this way before." I had seen the entrance to the passage before, and mentally marked it as a place I hoped to return to if I had the opportunity. We went through and scrambled down a breakdown slope into a red-floored passage, about five feet high and eight feet wide. Dead formations, dry as dust, covered the walls. Suddenly we heard voices above us.

"Hey!" yelled Luther. "Who's there?"

"Earl Thierry and Roy Charlton."

"This is Luther, where are you?"

"We're in the Turnpike Room at the back yelling down a crack in the floor. We were about to go down when we heard your voices."

"This is Brucker," I yelled. "We're looking for a pit. Do you know where it is?"

"Got one up here that's pretty deep, will that do?"

"No, this one's got a waterfall in it."

"No waterfall here," yelled Earl.

We moved on following several sets of footprints which Lou identified as those of the climbing party on their first attempt. But the footprints stopped after about a hundred feet when the sand floor changed to breakdown. Luther probed ahead into the darkness as we waited.

"Yaahhooo!" he screamed. "Virgin cave!" Luther Miller was in his element. He announced that we should give him ten minutes before starting to follow him, so he could check to see if the lead was blind or not. Then he squeezed around the side of a pile of dangerously loose breakdown.

In fifteen minutes he was back with a wild-eyed tale of the passage continuing onward. He had gone as far as another breakdown, then had come back for company.

Luther and I set a twenty-minute limit on our jaunt and left the others waiting in the passage eating candy bars. Soon we reached the second breakdown. Luther climbed up to look into a side passage while I picked my way through a small hole. The rocks were loose. The least pressure on some of them would have sealed me in, so I worked slowly and cautiously. Luther was behind me now, and I was happy to have company. Picking my way through this breakdown was a slow, deliberate process. My heart pounded with a hundred pictures of immense chambers, of wide underground avenues, of colorful formations, and of Floyd Collins who had been trapped in just such a place as this. I had no intention of suffering the same fate. One football-size rock blocked my progress. Above it and wedged against it were two other rocks which appeared to support the broken ceiling. Kicking against it would be asking for the same fate Collins had met, for he had kicked just such a stone, and it had fallen on his leg. After the rescuers had tunneled down to Floyd's body, they discovered the awful truth—the boulder pinning him down weighed only fifty-five pounds.

I could hear Luther breathing behind me, and could feel his eyes burning into my neck waiting for me to act or retreat. When I turned to look at him, he lay there expressionless, as if to say, "You're the boss." I beckoned for him to retreat about five feet, making a space for me should it become necessary to jump back in a hurry. I focused on the key rocks again, then slowly extended my left arm toward the topmost rock. I pushed it gently, feeling it move. The next time I pushed it harder and scrambled backward in case of an avalanche. All was quiet, terribly quiet, I thought.

Again I crept forward and extended my left arm, supporting myself on my right forearm. My tests had indicated that the top rock was not supporting the ceiling as I had originally believed. Several gloved fingers slipped into an invisible crack between the ceiling and the top of the rock. My hand curled around the stone and I eased it toward me in one continuous motion. So far, there were no signs of imminent collapse. I removed the second rock in the same manner, clearing a space so I could wrap both hands around the barrier rock. I lifted it off the floor, then slipped it back along my body where without warning it was taken by Luther. I had forgotten Luther existed during these anxious minutes, but he was there to help with no word from me. Luther could always be counted on as a "team" cave explorer who did more than his share. I was thankful he was there.

We squeezed through our new access hole and into a passage trending to the right. Fifteen minutes had passed, but we decided to press on since the passage was getting bigger. The red sand covered the floor again and in the distance we heard water falling from a great height into a pit.

We reached a spot in the passage where a second rock blocked our path. Beyond it we could shine our lights into a room, the dimensions of which we were unable to determine. "Shall we go on?" I asked.

"Let's go. Man, this is caving!" Luther had a hot gleam in his eye—the kind explorers get when they discover something new. I moved the rock and threw it into a shallow depression at the side of the passage. We squeezed through the narrow opening and stood up in a room where no man had ever stood before. It was thrilling . . . the cave primeval. We drank in the view. The room was about thirty-five feet long, and at the far end a passage led upward. To our left was a series of high domes, with pits beneath them. Several passages led off from the wall behind the pits. On our right, the brilliant red sand bank sloped up to within one foot of a rock ledge. Peering under it we saw a smooth expanse of sand and an equally smooth ceiling forming a pocket-like room perhaps sixty feet in diameter. The sound of falling water was loud now, apparently coming from the high passage straight ahead. We said little as we scampered about testing our discovery. A rock dropped into one pit took about three seconds to reach the bottom. Another pit was almost completely roofed over with breakdown, but rocks poked in holes there seemed to fall an eternity before hitting bottom. Gray-white limestone, colored differently from any we had yet seen in the cave made up the far wall. We walked over to it for a closer look and we saw what gave it its curious, rough appearance. The limestone was formed of millions of tiny fossils; crinoid stems, coral, miniature brachiopods. Fossils are quite often found in limestone, but in this instance there was no matrix for them. Water had dissolved the massive part, leaving only the fossils complete and whole. I had seen many crinoid buttons before, but never had I seen an entire stem of the prehistoric plant, complete with the hole down the center. We were hypnotized by the find. It was as though someone had taken a fine collection of Mississippian fossils and simply piled them here. They were not loose, but were cemented together at their edges, forming a bed at least nine feet thick.

Time . . . we had completely forgotten our twenty-minute deadline. We picked a name for the place—Lost Paradise—and vowed we'd be back to explore more thoroughly. The many passages leading off from this room, together with its fossil collection, had given us a kind of euphoria, and we wore big grins when we returned to the party.

We related excitedly the story of our find, and apologized for staying forty minutes instead of the twenty we had agreed upon. But the group had not been sitting idly by. Lou had found the passage leading to the pit we had been sent to cross. Following the others we followed a sewer-type passage on all fours. The floor began to slope downward deeper, but we worked along the sides, so in effect we were chimneying forward at the top of what was now a deep canyon. It curved to the left as we went. In about ten minutes we heard falling water ahead and soon came to the brink of a large pit. We chimneyed down to the bottom of the canyon at a point where we had firm footing.

"There's where we're trying to go," said Lou, shining his flashlight into a passage continuing onward on the other side of the pit. Thirty feet of open air separated us from the passage. We estimated the distance to the bottom of the pit to be about fifty feet, and the distance to the ceiling at about forty. By leaning out into the pit and looking up to the right, we saw a stream of water pouring out of a passage near the ceiling. Its fire-hose volume curved gracefully into the abyss, splattering itself on the black rocks below.

We held a conference over a few candy bars to decide the best way to tackle the problem.

Lou recounted the experiences of the previous day's attempt. He and Audrey had rigged a piton belay and Lou had rappelled to the bottom. Crossing the pit he had been able to scale the muddy wall to within about eight feet of the passage, but had turned back for lack of a second man to rope down and cross, so Lou could stand on his shoulders. The party had felt that putting two people in the pit would have jeopardized their chances of climbing back out. He suggested that on this trip we put two men down, since we now had a larger party. It looked difficult and dangerous. The consensus was that it would be wise to reach the bottom again and then make a decision from there. Lou began to uncoil the rope.

An idea struck me. Didn't the passage into Lost Paradise trend to the right? And hadn't the passage we had followed into the pit led us to the left? If my calculations were accurate, we would be able to get to the other side of the pit simply by following the route into Lost Paradise and continuing along the high passage until it completed the link. We *had* heard water falling ahead of us in Lost Paradise.

I explained the idea to the others and they agreed that it would be better to try that route first. It involved less risk; in exploring Crystal Cave one doesn't do rock climbing for the fun of it. Explorers use rock-climbing techniques to get some place, and don't resort to them unless all other possible routes have been exhausted. I told the others to wait, and if all went well, I'd be shining my light at them from the far side of the pit in about 20 minutes.

Lost Paradise had lost none of its original aura when I entered it again. I moved to the far side to investigate a possible route up to the high passage, finding a splendid place to chimney up. The top was muddy as I squirmed around to get into crawling position. The sharp-edged fossils, beautiful as they were, tore my coveralls in several places. I pushed on toward the sound of the water, through pools of ooze and past fresh breakdown. Water was actively doing its work in this part of the cave. My light sputtered for an instant, so I refilled it. A wave of loneliness swept over me. I heard nothing but the splatter of water ahead and, nearer, the drip of water into a dozen pools. I shivered for a moment, debating whether or not to go back, but the loudness of the water bolstered me with the thought that I would soon see the others.

In a few minutes a hole loomed ahead of me with a blast of air pouring through it. I thrust my head out and looked straight down into the yawning depths of a pit. Was this the right one? I couldn't tell. The most obvious sign would have been the others in the passage on the far side. But they weren't there! Perhaps they had returned to find me, since I had been gone over forty-five minutes. Below the hole from which my head and shoulders protruded was a narrow ledge, about a foot wide. The wall offered plenty of handholds. The ledge led to a high upper passage on the right, one whose view would have been obstructed from a vantage point on the other side. I saw a way of getting from the high right-hand passage into the one directly opposite me. If I wanted to catch up with the others, it would be easier to cross the pit rather than to retrace my steps. I decided to follow this course of action, planning to complete the circle.

I pulled my head out of the hole and jockeyed my body around in the tight muddy room in which I was lying. I backed up on my elbows, working my feet out into the emptiness of the pit, then I lowered them until they touched the ledge. A rock on the ledge went plummeting down, smashing with a sickening crash on the bottom. I pulled my trunk into the pit and started to inch my way along the ledge to the high passage.

For an instant my heart stopped. Just as I was moving my right foot a three-foot chunk of the ledge broke loose, hitting with the sound of an explosion below. I looked at the route over which I had just come, now only a sheer wall with wisps of dust hovering over it. I was trapped! My heart pounded like a sledge hammer. What if the part of the ledge I was now standing on let go? For a split-second I saw myself at the bottom of the pit, crumpled in a heap. I quickly continued moving, making sure of every handhold before putting any weight on the ledge. I sat down panting when I reached safety in the mouth of the high passage. Now I was in peril! I couldn't return if I wanted to. The others would be after me soon, I thought, but what could they do? I had been a fool to break the first rule of caving: *Never explore alone.* The passage I was sitting in was about twenty feet high; that is, I *thought* it was a passage.

When I summoned wind enough to examine my surroundings I found that my passage was really a bridge between two pits! And the second looked as formidable as the first. I returned to the mouth of the passage and thought. What could I do? I turned on my flashlight, happy that it worked after being dragged through mud and water on the way in. I played the beam along the walls of the pit, looking for any route back. There appeared to be none. The collapsed ledge would be impassable without help, and if help were to come, there literally would be no room to maneuver in the constricted passage from which I had come. On the opposite wall of the pit my light picked up a red flash. I moved the beam back again and discovered a strip of reflective tape stuck to the wall. Beside it in carbide was the inscription "Pit #3." Obviously someone on the expedition had crawled out of the passage in that opposite wall and marked the pit so it could be referred to in the future. But getting down to that passage where he had emerged was a problem. Only one route held promise. It lay in working one's way down the top of a huge boulder, tipped at a precarious angle into the mouth of the pit. There were no handholds, and the boulder was covered with dry mud, making matters worse. If friction failed to hold me when I attempted to cross it, I would slide down the boulder and plunge into the pit. The only chance of stopping myself would be the last chance, the one on the bottom, and that thought didn't appeal to me.

Straight ahead my light picked up another bridge, just like the one on which I was sitting, and beyond that, the blackness of another pit. What I had stumbled into was a wide pit-canyon series that continued as far as my light could reach in both directions. What a potential for exploration, but then, would I be able to tell anyone about it? I sat down to wait for help. When help did not arrive, I examined the boulder again. Perhaps with some careful work I could make it to the side passage. The top surface of the boulder tipped at an angle of about 30 degrees into the pit. I climbed on the top near the back, holding on the edge of it with my hands. It was slippery, just as slippery as it had looked the first time. With one hand I picked up a sharp rock and scraped the mud from between my rubber shoe cleats; I would need all the traction I could get. The cleats sank into the mud when I placed my feet flat on the surface. To increase the area of holding friction, I lay flat on my back in the mud and spread my arms wide.

One path would lead me to safety. It would be a slow and deliberate edging downward for a distance of about ten feet to a point about two thirds of the way to the bottom lip of the rock. At this point I would have to stop my downward progress and work my way to the right for about five feet, then swing my legs off the boulder on to a small breakdown slope at the mouth of the side passage. Only then would I be out of danger.

Slowly I moved my left foot a few inches, then my right foot. With my arms I pushed my body downward after securing the best resemblance to a foothold I could obtain. Each time I repeated the process I advanced a few inches toward the passage, and toward the bottom edge of the boulder with the yawning pit below. Three feet to go before starting to move sideways . . . now two feet . . . then the cave was plunged into blackness!

Roaring water thundered in my ears. A draft had blown my carbide lamp out, leaving me stranded a scant three feet from a one-way shaft down. Petrified with fear I lay there breathing deeply. Moving would be folly, for the slide was hard enough to cling to when I could see, let alone in blindness. My flashlight! That was my hope! Getting to it involved loosening a precious handhold, but I took the chance. I edged my hand toward my waistband where my flashlight dangled. I found it after groping and eased on the switch with my thumb, then moved my arm back into position. The light was marvelous, pouring warmth into the whole cave, and into me.

In another five minutes, I reached the safety of the passage. All strength left my body. I slumped back into the safety of the passage, thankful not to see the bottom of the pit at close range.

My reflections were interrupted by a faint voice calling to me. It was Luther! My heart quickened. I yelled an answer. He came closer and in five minutes I saw his bobbing light in the hole from which I had entered the pit.

"Come on back!" he yelled.

"I can't." I shouted back.

"Come on back," he repeated.

"I got here by crossing a ledge. The ledge collapsed and now I can't get back."

"Where are you?" he asked.

"I don't know. I've never been here before." It was difficult shouting over the din of the water.

"Can you get back to Camp One?"

"I don't know," I roared. "Come closer so you can hear me!"

The light grew stronger as he approached the hole. "You don't recognize anything?"

"Nothing at all. Did you ever hear of pit number three?"

"No, not unless it has another name." Luther knew as much about the cave as anyone in it, including Jim Dyer, and I had counted on his knowing just about where I was. I shouted over the details of my surroundings.

"Doesn't ring a bell," he said. "I think you're in something brand new."

"There are tracks here," I shouted. "Someone's been here before."

"Follow them. They may take you to something you recognize," said Luther. "If you don't get back to Camp One in forty-five minutes, we'll come looking for you!"

I said goodbye and watched his light trail off into blackness, then I followed a fresh set of footprints along a low passage leading away from the pit. I squeezed through a small hole into a larger passage and stood up, breathing easier now. Directly over the passage where I had come out, there was a carbide dot with a circle around it, the symbol for unexplored cave. One other person had entered it before. His footprints and marking had led me to safety. This was one time I was happy to take second honors in finding a new part of the cave.

I set up a cairn to mark the spot by piling up loose rocks, so we could come back again and explore the canyon series more thoroughly. The footprints led down the passage and in a minute or two I recognized where I was. This was quite close to the Lost Passage, and I might even beat the others back to camp. I entered the spring basin, crouched through the muddy tube, and jumped off the ledge into Floyd's Kitchen. Ten minutes later I walked into camp.

Luther and the others pulled into camp forty-five minutes later, astounded to see me already eating. We wrote a brief account of our doings in the log, and talked to Skeets and Phil Harsham about our find. Later I learned that the unknown explorer who had so kindly marked Pit #3 was Roy Charlton. He and Earl Thierry had stumbled into it only one hour or so before I had, and they too were eager to go back.

I went to the spring to wash, then headed for a warm sleeping bag.

We had found a brand-new cave; a totally untapped area for exploration with excellent possibilities for discovery. But the die was already cast. Lost Paradise would await the return of another expedition, for the order had been given to establish Camp Two in another part of the cave.

Part III

ANOTHER BARRIER AND CAVES BEYOND

by Joe Lawrence, Jr.

17

CAMP TWO

LATE WEDNESDAY NIGHT, prior to Brucker's return from the Lost Paradise, Bill Austin and I sat in the cave office planning tomorrow's operations for the expedition. We had just received startling news from Jim Dyer at Camp One. Bob Handley's exploring party had returned with an exciting tale of discovery.

It had been early Wednesday morning when Bob Handley had departed from Camp One to explore some virgin leads in the X Pit area. From my sleeping bag near the phone I had watched the bobbing file of carbide lights recede over the rolling sand floor of Floyd's Lost Passage. It was a large group of seven that Handley led, for it was really two parties. A scientific party wanted to collect blind fish in the bottom of X Pit. Handley was in charge of an exploring party. Bob Handley was the only one of the group who had been to the bottom of X Pit, so I told him to lead both parties that far.

Several years had gone by since Handley had been to the bottom of X Pit, and time dulls one's memory of Crystal Cave. Getting to the top of X Pit was easy; this was a well-known trail. But as Handley picked his way down through the seldom-traveled, intersecting passages, he was less sure of himself. He came to a junction where he had three choices. One, a crawl, seemed more familiar to him than the others. Handley crawled on his right side now; days of long crawling had made his left knee quite tender. He had gone fifteen feet when he decided the passage just didn't look right and told the party to back out. While one of the biologists tried one lead, Handley tried another. Both were wrong. Bob went back into the crawlway, his first choice. It proved to be the right one after all, and soon

When supplies arrived in Camp One packed in newspaper, Bob Handley was quick to take advantage of the situation. He catches up on the outside. *Photo by Robert Halmi.*

the explorers and biologists stood on a balcony looking into the blackness of X Pit.

The rest of the route was easy for Handley. A tight squeeze to the left led to another balcony just thirty feet above the floor of the pit. A crawl around the side of the pit on a ledge brought them closer to the bottom. Handley tied a rope to a projecting rock and started the party down. Then one person balked. He didn't feel he could make the descent on the rope safely, and he said so. This was exactly what he should have done. Because most of the expedition members refused to undertake anything they considered dangerous or beyond their capabilities, we had avoided serious accidents so far. Handley sent the uneasy explorer back to Camp One with a companion.

At the bottom of the pit, two members of the biology party flattened themselves on the rocky bank of a pool and peered into the depths waiting for some hapless blind cave fish to swim out from under a ledge into the revealing glare of a carbide lamp. Bob Handley and the two marines, Kendall and Longsworth, departed to explore.

The explorers climbed into a side canyon that went up. It was a deceptive climb because the slick, heavy mud on the walls made the ascent more difficult than it appeared. The mud must have been deposited by periodic flooding of X Pit, for the floor of the Pit was quite close to the water table, apparently, since the water level in it rose and fell with fluctuations in the Green River water level. As soon as they had gotten above the mud, climbing was easier. Now they were in a canyon with a level, gravel floor. Handley looked carefully for telltale scuff marks on the floor, but there were none. They were in virgin cave!

By their calculations, this passage lay below any others in the area that had been explored. They loped along now, following the canyon south until the floor dropped steeply downward. They could hear the roar of the water below, so they climbed downward. The canyon was high here, and the walls as narrow. They estimated the gloomy ceiling eighty-five feet above them. It was a water-cut canyon with smoothly carved walls, and here and there projections of harder rock jutted out casting fantastic shadows on the walls as they moved between them.

Their descent brought the explorers to mud, then to a stream that flowed through an intersecting canyon. They looked in surprise at the rush of a solid stream of water fully six inches in diameter gushing down from above. They turned right and proceeded upstream, climbing falls and straddling rapids. Then they came into a small room with two names on the wall—Dyer and Austin. They were back in the explored part of the cave and near the C-3 Waterfall—the largest known fall in the cave and the one the press representatives had been taken to see on the day before by a different route. The waterfall gained its name from the first three initials in Collins Crystal Cave.

The C-3 Waterfall lay at one end of the Waterfall Trail. From here the trail led to the top of X Pit and on to Bottomless Pit. What Handley had found was a lower-level passage beneath the Waterfall Trail. We already knew that above the Waterfall Trail was the Crawlway. And above the Crawlway were Scotchman's Trap and the tourist trails. The cave was stacked deep here. It was the twisting, crossing and intersecting passages on many levels that made the cave so confusing. Explorers would cross under and over passages without knowing it. The constant turning made it impossible to maintain a sense of direction without a compass.

Occasionally explorers would break through from virgin cave into a passage they knew. Then another piece of the puzzle would fall into place.

This is what Handley had just done. It is what Lehrberger had done earlier in the week when he found a new link between the B Trail and the Bogardus Waterfall Trail. Slowly, we were shaking loose some of Crystal's secrets.

When Bob Handley reported his discovery and described the many side passages he had left unexplored, Jim realized that here was a whole new area for exploration. He turned the crank on the Camp One phone and asked for Austin. I was with Bill when the telephone rang in the winter office. By the expression on his face I knew something important was going on. I pressed my ear next to his and was able to hear Jim relate the details of Handley's findings.

Bill Austin's recommendation to me came the minute he hung up the receiver. There was a good camp site with plenty of drinking water at the C-3 Waterfall. We ought to establish a camp there so that we could push Handley's canyon discovery to the limit. I hesitated; we had had only one exploring party in the area. We couldn't be sure that there was enough work there to justify setting up a second camp. I rose and paced back and forth across the office floor while Austin reminded me that we had pulled out of the Bogardus-B Trail area and that Mud Avenue had been abandoned for further work. Since we were now probing in all directions from Camp One, we had to seize the best potential discovered up to this hour. The probing parties had found many areas of cave that needed more work, but up to that time Handley's discovery seemed the most promising of all. The argument was convincing; I decided to establish Camp Two at the C-3 Waterfall.

We pored over the roster to decide what personnel to transfer. Austin and I had both noticed that fatigue was very evident in those who had been in the cave a long time. We must pull out those who were worn down and send in fresh explorers from the Support Group to replace them. And we must pick our best explorers for Camp Two. We made three lists.

Those in Camp One who needed to come out for rest were Audrey Blakesley, Lou Lutz, Roger McClure, and Roger Brucker. As we made this list I remembered Brucker as I had last seen him. His face was covered with a heavy beard. He had tired red eyes set in deep hollows, and he stared vacantly at times. I had observed him shifting his weight from one foot to the other and had noted his lapses into silence. I knew how these four people felt, because I had felt this kind of fatigue that was hanging over the Camp One explorers like a blanket. I remembered how, earlier in the week, I had stood on the smooth floor of Camp One talking to a returning survey party when, for no reason at all, I had lost my balance. I had recovered without falling by taking a quick step backward, but someone had said, "Joe, you should go topside and get some rest."

"No, I feel all right," I had said. Then I felt something hot pressing against my hand. Ida was trying to hand me a cup of coffee.

"No, thanks, Ida. I really don't want any coffee now." But already the

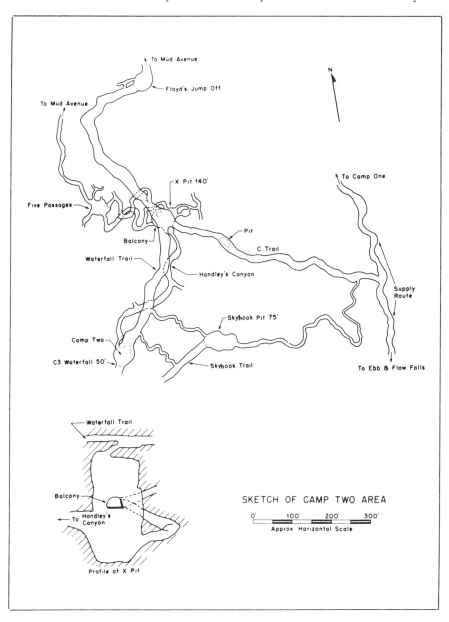

SKETCH OF CAMP TWO AREA

0'   100'   200'   300'

Approx Horizontal Scale

cup was in my hand; I was automatically pressing it to my lips, and the warm refreshing liquid tasted wonderfully good.

It was necessary that those "cave happy" with fatigue come out; otherwise, we would run a serious risk of an accident. Austin rang Camp One. When he reached Dyer he read the names of the four to be sent out. There would have been more evacuated, but in the past couple of days several people in Camp One had voluntarily requested to be sent out for rest. Brucker and Miller, I learned, had not yet returned from their attempt to climb Frenchman's Pit.

We scanned the roster for our best explorers who were still in good physical condition. Bob Handley would lead them to the C-3 Waterfall and establish Camp Two. Although he should have been tired, he seemed to be bursting with energy. Bernie Wilt, a bricklayer, and our two marines had proved themselves indestructible on fast-moving support parties. They appeared in good shape, and they deserved a crack at new cave, so we as-

Earl Thierry sleeps soundly between interruptions. *Photo by Robert Halmi.*

signed them to Camp Two. Earl Thierry was showing few signs of fatigue, although I learned he had just gone to sleep. He would go with Handley.

"What about Charlton?" Austin asked.

"He looked eager when I left Camp One this morning, although this is his fifth day in the cave."

"We could have him help set up Camp Two, then come out with the next support party," said Austin.

"That sounds all right," I said, then more orders were issued to Dyer.

The next list we prepared was of Base Camp personnel to be transferred into the cave. This went to Stephenson. Then to Alden Snell went a list of supplies to be sent to Camp Two. Bob Lutz was asked to set up a telephone in Camp Two as soon as he could.

Thursday morning, the plans became action. Lehrberger led a party from Base Camp with wire to be laid to Camp Two. Stephenson and two interns departed with supplies for the new underground base.

Bob Handley carefully supervises the packing of supplies for Camp Two. His camp will be self-sufficient, so he sees to it that his group has all of the essentials. *Photo by Robert Halmi.*

Back in Camp One Earl Thierry and Roy Charlton had crawled into their sleeping bags an hour before the Camp Two personnel were scheduled to depart. No sooner had Earl dropped off to sleep than someone waked him and told him he wanted his sleeping bag.

"What's the matter? I just got in," complained Earl.

"The tag on your bed says you went to sleep at 9:30. That was over twelve hours ago."

"That little tag says 9:30 A.M."

A flashlight was hastily played on the bed tag. "Oh, sorry. . . ." and Earl's tormentor disappeared into the darkness.

Handley's group signs out in the Camp One log, then files off down the passage as Jim Dyer looks on. *Photo by Robert Halmi.*

Earl dropped off to sleep again, but was soon awakened by Bob Handley. "You're on the list to go to Camp Two. We're pulling out in half an hour."

"Camp Two?"

"Yeah."

"Where's that?"

"C-3 Waterfall. Let's go."

"I don't think I should without sleep."

Then Earl saw Roy was already up and pulling on his trousers. Reluctantly, Earl crawled out of the sleeping bag and got dressed.

Loaded with sleeping bags and a day's supply of food, Bob Handley's explorers filed off toward Mud Avenue. It was Thursday morning, a very dark morning about two hundred fifty feet below the surface.

Shortly after the Camp Two team pulled out, Dyer relayed to me the word of Brucker's and Miller's discovery of Lost Paradise. He reported to me in detail on the potential that lay in the area, but I had already committed the expedition to another course of action. Calling the explorers back now, especially without phone communications established, would mean a complete loss of at least twelve hours of exploring time. I felt that in the forty-eight working hours remaining, we should spend as little time as possible moving into position to explore.

What if the Lost Paradise led to even bigger caves beyond? And what if Handley's side leads pinched out? I caught myself debating the wisdom of my decision. As a rationalization, I supposed the opposite to be true; that Handley's canyon led to more cave, while the Lost Paradise pinched out. But I couldn't be sure. Omniscience is the greatest single gift an expedition leader can have: I felt poor in that thought until I realized that the hindsight I would have at the end of the week might be an acceptable substitute.

At 3:40 P.M. on Thursday I entered with Al Mueller and Russ Gurnee to inspect Camp Two. We carried food, gasoline, and cooking gear, together with two hot beef-roasts wrapped in aluminum foil and layers of newspaper to keep them warm. They reposed neatly in a Gurnee can along with other supplies.

In the tight Crawlway there was real danger that the gasoline we were carrying might explode. We carried the gasoline in several one-gallon cans packed together inside another Gurnee can, but even with this careful packing there was enough leakage around the caps to engulf us in gas fumes. We had been warned about this early in the week by support teams who reported narrowly escaping being blown up on their first trips carrying gasoline. They were in constant fear of their carbide lamp flames ending the expedition for them then and there. For this reason, all subsequent teams carrying gasoline had been issued miners' safety lamps run by a battery attached to their belts. Two of us wore these electric lamps, but the supply had been short and Al Mueller carried his carbide lamp as usual. We asked him to keep his distance, so he trailed behind like a leper.

On one occasion we saw the approaching carbide lamps of a returning support party.

"Stop, we have gasoline!"

They froze. Many of the support personnel had been in gas-carrying parties before. They knew how the fumes hung like a plague about such parties, and now these support teams without the safety of electric lights were confronted with a gasoline party. They stopped and waited. We advanced to a wide place in the passage, moving the gasoline canister as far to the left as possible. Then cautiously but rapidly the carbide-lighted support party scampered by, pressing the right wall. As they disappeared behind us our tension left. We snaked the gasoline canister back into the trail and proceeded.

As we passed the Keyhole we found the battered nine-and-one-half-inch oven where I had left it. We added it to our load.

We arrived at the junction where the Camp One trail continued on and the Camp Two trail led down through a small hole in the floor. The end of the Camp Two telephone wires lay there unconnected. They had been laid from this junction to Camp Two just a few hours before by Lehrberger's wire party. The other four conductors reached onward to Camp One, carrying two telephone circuits that terminated in the Lost Passage. Since four conductors can carry three circuits provided one is a phantom, Bob Lutz had planned weeks before the expedition to provide two circuits to Camp One and one circuit to a possible second camp. Before I had left Base Camp, I had asked Bob Lutz how to connect the two conductors leading to Camp Two. "Tie them in any way you can, just so you can rouse us upstairs," he had said; "I'll straighten them out later."

I stripped the insulation from two of the Camp One wires and spliced on the Camp Two wires. Then we moved down the C Trail following the Camp Two wires.

When we reached X Pit, we found a telephone. I spliced it into the line and turned the crank. When the Base Camp operator answered, I asked for Bob Lutz and soon told him that the Camp Two telephone line was connected.

As we proceeded to Camp Two we passed Bob Handley and the tireless Luther Miller. They were working on the new telephone line, but Luther paused long enough to offer me a side trip to Lost Paradise. His eyes gleamed when he talked about what he had seen there. Regretfully, I had to decline.

We crawled along the soft dirt of a low wide passage to a crack in the floor, the top of a canyon of unknown depth. We reached a point where the canyon widened and where an arrow smoked on a boulder pointed down into the slot. Also written with the smoke of a carbide lamp was the notation "Camp #2." We climbed downward into the canyon by bracing our hands and feet on the walls. The precipitous route led us into the base of a cylindrical shaft ten feet in diameter, where Camp Two was spread out on the floor. It was deserted. A stove, a supply of food and a sleeping bag lay nearby. Leftover oatmeal rested sluggishly in the bottom of a cold pot.

One side of the shaft was open. Here, we could look down into the "basement" of Camp Two and see a sparkling torrent of water racing toward us and under us. We could hear the roar upstream of the C-3 Waterfall. To me, this was one of the most beautiful spots in the cave. Golden-brown walls studded with dozens of shadow-casting ledges lent the place an air of mystery, heightened by the sound of water plunging down a deep shaft.

Sleeping on the "20th floor" at Camp Two is difficult because of the roar of the waterfall and the chatter of passing supply parties. *Photo by Robert Halmi.*

We put down our loads, then Russ Gurnee, the sheet-metal contractor, took a closer look at the mangled oven. When he saw in a clearer light what I had done to force it through the Keyhole, he sat down and almost cried.

"What *did* you do to it?" he asked. "How could you?"

"I compressed it," I said. "Now it's up to you to fix it."

He shook his head from side to side and emitted a low whistle. "I don't know . . . I'll need tools."

"As a tin-knocker of good reputation and builder of the famed Gurnee can, try to figure out something," I said.

He asked for a hammer. We had none. He begged for a pair of pliers.

There were none in Camp Two. Russ crawled into a corner where he picked up a rock, eyed it carefully, then began to pound the oven with a clanging rivaling that which I had made the day before at the Keyhole. When he had finished, it looked recognizable considering what it had been through. The door would even open and close; only the thermometer had been lost.

Russ set the oven up on the stove and soon had biscuits baking. I unpacked one of the beef-roasts. The newspapers and aluminum foil had kept it warm, providing us a delicious supper of hot biscuits and roast beef.

Phil Smith, Luther Miller, and Bill Stephenson joined us in camp after depositing a load of supplies above. In a little while, Mueller, Gurnee, and Stephenson departed together.

Simultaneously, the two marine lieutenants arrived in camp from an exploring mission. They had been probing side leads that started near the top of the waterfall. All of them led into narrow canyons too small to travel

The Camp Two kitchen, nestled at the bottom of a pit, includes a stock of food, a gasoline stove, and a battered biscuit oven. *Photo by Robert Halmi.*

Lawrence climbs out of the Camp Two kitchen carrying a beef roast for Camp One in a Gurnee can. This climb is typical of the many that are made as a matter of routine by Crystal Cave explorers when going from one place to another. *Photo by Robert Halmi.*

after a hundred feet or less, except for one very wet crawl that the marines didn't want to explore without a change of clothes, and so they had returned to camp seeking companionship and food. I could see exhaustion reflected in their eyes, and the eagerness they displayed early in the week was replaced by a kind of lethargy unusual for explorers located so close to potential discovery.

Luther Miller and I logged out for Camp One. With one of the beef-roasts bouncing along behind in a Gurnee can, we sped over the Waterfall Trail.

Bob Handley sent the weary marines to bed. He dispatched Charlton and Thierry to explore near Ebb and Flow Falls. Then Handley dragged a sleeping bag over beside the telephone. It was a temptation to run a camp from a sleeping bag by a telephone. Handley tried it. He made several routine calls; then he put through a long distance call to his girl friend in Charleston, West Virginia. She was thrilled to be able to talk to Bob while he was in the depths of Crystal Cave. Then Handley settled back to go to sleep. The phone rang, it was a call from Base Camp. When he had finished he tried again to sleep, but was disturbed once more by the telephone. When he had finished this call, Handley rang the operator, "I'm trying to get some sleep down here. Don't put any calls through to Camp Two except in an emergency."

Then Handley slept. He slept for two hours before he was awakened again by the ringing telephone. In Base Camp the operators had changed shifts and the retiring operator had failed to relay Handley's instructions. Muttering to himself, Bob Handley dragged his sleeping bag out of earshot of the telephone.

Lehrberger and Welsh directed their first effort at searching out the source of water for the C-3 Waterfall. They chimneyed their way out of Camp Two kitchen, then moved past the silent forms of sleeping men to a narrow ledge leading behind the falls. Their route led them under the arc of water itself, but they remained reasonably dry in spite of some spatters reaching them. They eyed the three-foot-by-three-foot passage cautiously. Lehrberger boosted himself into it, shouting back that the passage looked passable but that he was in water up to his ankles. He told Welsh to wait before coming after him; there was no reason for two getting wet if the passage pinched down.

In a few minutes Lehrberger climbed back down into the main passage, his clothes dripping with water. He described how the floor had slowly risen to meet the ceiling, and how he had been able to crouch part of the way, but soon was on his hands and knees. When this became too tight, he had dropped to his belly, pulling his body upstream by handholds on the wall. The icy water had forced him to turn back at a place where only his nose and eyes were above water. "I would have gone on," he said, "but I didn't want to get wet." He was shivering now, so Welsh suggested they move on.

They pressed beyond the waterfall into three leads, all of which were blocked after a short distance by sandstone breakdown. They reasoned that they were running into the bottom of a surface sinkhole, and they thought that a map would prove it.

Phil Smith and Earl Thierry teamed up to explore side leads off Handley's Canyon, but those that they could reach went nowhere, and the others up higher were denied them because their party was not big enough to attempt major rock climbing.

Rejoined later by Charlton, they moved out to look over the Skyhook area, a system of passages reached by crawling along a narrow ledge in the canyon under Ebb and Flow Falls. Below them, a stream gurgled over the rocks. When Smith in the lead found the ledge narrowing, he shifted his feet over to the opposite wall, making a human bridge across the chasm. The party moved sideways in this fashion for about ten minutes until Smith yelled to the others that the passage was getting wider, and he was afraid he had extended himself to the limits of his reach. Phil Smith was clearly in a predicament that only he could remedy. His left and right hands were braced on a four-inch ledge. Five feet across the yawning passage his feet were pressed against a rough wall, holding him only through the sideways force he could exert with his arched body. A fifty-foot drop yawned below. The others couldn't help him back to safety, for they too were extended in similar positions, although their bodies spanned the passage at a point where the walls were closer together. If they tried to move out to him, there would be no room to conduct a rescue.

Smith's breath came in short, deep gasps, the unmistakable sound of fear. Charlton knew that Phil was an experienced man, and seldom, if ever, had he shown signs of being terrified. Knowing something had to be done, Charlton did the only thing he could; he began to talk, calmly and casually. "You're all right, Phil. . . . Catch your breath before you try anything. . . . Move your right hand toward your left. That's it. . . . Now move the left one. . . . That's good. . . . Now try moving your feet. . . . Coming fine. . . . Keep it up." In this way he somehow summoned forth a new reservoir of strength within Phil's body. Smith and the others carefully worked their way back the passage to safety. As they rested, Charlton and Thierry watched Smith, who was tired to the point of trembling.

After a rest, they back-tracked to a place where they could chimney down to another ledge. This led them in about ten minutes to a natural breakdown bridge of boulders which had jammed across the passage.

"They call this the Skyhook," said Roy. "Don't ask me why." Smith remembered the story Jim Dyer had told the first time he had visited the cave back in November. Jim had said that Luther Miller once found a big cave passage near the Skyhook. The group shined their headlamps into the gloom wondering what lay beyond.

In another few minutes they were sleepily eyeing a high passage, accessible only by climbing up rotted-rock hand- and footholds. Smith touched one of the projections and watched it plummet into the water below. The tired explorers looked at each other without speaking, then looked upward at the passage. The temptation to try to reach it was great, but even their fatigue-numbed minds told them that the risk was not worth it. Perhaps when they had rested they would come back and try it. They gave up the high passage and the canyon stretching onward, forsaking both for a hot meal and warm sleeping bags.

Handley listened to the experiences of still another party a few hours later. They had gone to Five Passages to solve once and for all the riddle of where all seven passages led. Splitting into two groups, they had entered two passages simultaneously. Both passages were virgin, but that was all they could say for certain about them. They moved up and down, confronted at every turn by a junction of passages so similar in shape that it was impossible to tell which was the main passage. They plunged on, arbitrarily choosing their route from dozens that offered access. Fortunately, they had marked arrows on the wall, for after two hours of exploring with no change in the character of the passages, they decided to retreat from this "bowl of spaghetti." At one junction a party overlooked an arrow placed on a wall in such a position that only an incoming party could see it. After fifteen minutes of crawling on hands and knees without seeing an arrow they decided they had made the wrong turn. They followed their marks in the sand floor as eagerly as bloodhounds on a scent. They soon found the cause of the error and resolved to be more careful in the future.

Friday afternoon when Bob Handley awoke, the gloom of the cave seemed heavier. Everywhere he sent his explorers he got discouraging reports—meaningless crawlways that twisted and intersected, leads choked with breakdown, walls of rotten rock too muddy to climb, wet passages and no dry clothes. Handley looked at one of his returning exploring parties. Their heads hung with fatigue; their eyes blinked slowly; their conversation was sparse, sometimes meaningless. He looked at those crawling out of sleeping bags. They were supposed to be rested but they had gotten very little sleep. It was they he must send out to seek new discoveries —new discoveries in this hole. Was there anything new down here? After a week it all looked alike. Wasn't it all just rock, mud, and water? Handley, himself, had gotten very little sleep in the last forty hours.

He knew that Camp Two's productive hours were numbered. Keeping up this killing pace of exploring would be impossible. Early in the week Charlton and Thierry had thrown their energy into the assault on B Trail, and they were now paying for their discoveries. Smith, Wilt, and the marines had bulldozed supplies through the Crawlway in the early days of the expedition. They could not recuperate from this overnight. Handley realized now that what kept explorers going was sheer enthusiasm and zeal to tread in the unknown. With enthusiasm and zeal quenched by fatigue, there would be no more exploration from Camp Two. He knew it but he hated to admit it, even to himself.

As Handley knelt to write in the log he winced. His left knee, now infected, was worse today. Crystal Cave's implacable endurance limit had moved in. It lingered just beyond the circle of light that was Camp Two. Then he wrote, "1700 hr. Feb. 19. Handley up to check feasibility of breaking Camp #2—Charlton is dead tired, can't sleep—Handley has infected knee —Kendall and Longsworth want to leave." Crystal Cave clung jealously to some of her secrets.

A tired crew nibble rations in the Camp Two kitchen and dejectedly discuss their failure to find anything new in their assigned area. *Photo by Glen Shoptaugh.*

18

NO GREEN WATER

ABOUT NOON ON THE 18TH Roger Brucker was shaken by a silent figure. "Brucker, get dressed," it whispered. "You're scheduled to lead a party out of the cave to pick up supplies." He dressed hurriedly and went down the passage to Camp One to get something to eat.

Roger McClure, Audrey Blakesley, and Lou Lutz were eating. They would be in the party, and already had gathered together the bundles they would take out. It wasn't much of a load, they remarked, as they stuffed rations in their sacks and signed out in the log book.

"Tell them at Base Camp this is going to be a leisurely trip," said McClure. "I want to get some pictures on the way out."

Each person had mixed feelings when he started. They needed a rest on the surface, but there was the gnawing feeling of leaving an unsolved puzzle. The fact that their turn had come to carry in supplies, however, gave them no choice in the matter. It was only fair that the supply teams be given a chance at the thrill of exploration.

The trip out covered the old familiar ground of the supply route, now trampled flat by the dozens of pack trains thundering over it since the start of the expedition. Gay red arrows pointed the way at misleading junctions. Rog McClure took pictures at Ebb and Flow Falls. About 5:00 P.M. they reached the start of the Crawlway, where they sat down to eat.

"You know, a good steak dinner would taste good right now," said Audrey with a hungry smile.

"I'll second that," added McClure. Lou and Roger agreed. Perhaps when they reached the outside they could drive to civilization and eat a large T-bone. The prospect spurred them on.

Lutz, Blakesley, and Brucker, scurry along a well-traveled supply route to the surface. *Photo by Roger McClure.*

About 7:30 P.M. they phoned Base Camp from the Trap to let the topside crew know they were on the way out. Did they want anything to eat? No, they would take care of that.

The commercial part of the cave seemed cold and deserted to them as they walked back. Nobody was laying wire. The lights were off. No flash-bulbs were popping. Silence dominated. They talked about what would be going on back at Camp One; explorers would be wandering in and out with stories of new and bigger finds. And somewhere in Camp Two, carbide lamps would be shedding light on more explorers, weary with fatigue. Thin strands of wire which they were now following connected those deep in the cave with civilization and reality.

They each grabbed an apple from the bushel basket at the cave door, then climbed the stairs into the brilliant night. The moon was out, and the deep blue sky was veiled in places by patches of buttermilk clouds. Black trees were silhouetted against the luminous blue, a brisk February wind bending them back and forth. To them, it was electric. For almost a week they had been without the sight of blue coloring which is taken for granted outside. In the cave, all was brown, red, tan, white, or gray. There was no blue, and the novelty of this realization hypnotized them.

The party trudged up to the kitchen where the lights were blazing. They didn't eat, however, for they were holding out for the steak dinner they had promised themselves. Showering proved to be a wonderful experience, one of those almost forgotten luxuries in the sleepless nights and days below. It felt good to be clean again. Within an hour they had left Crystal Cave and had driven away to a restaurant, miles from the expedition. A dozen customers stared at them as they entered. It must have been the beards they wore, for they had been growing for a full week and the explorers had not thought to shave them off. A steak dinner followed by the comfort of a Base Camp tent made the day perfect for them.

Seeing the sun for the first time in a week can be an awesome experience also. Brucker had to squint as he crawled out of his tent into the blaze of daylight and sauntered over to the kitchen. Roger McClure was already eating.

"They won't be ready for us to take in a load until this afternoon," said Roger. "Why don't we take the altimeters and run an elevation study?"

"That's a good idea. We can take all the altimeters to the nearest bench mark and set them for the correct elevation," said Brucker.

The expedition had four altimeters. The first one was a large and highly sensitive American Paulin System instrument which would read to the nearest foot; it was designed as a surveying altimeter. There were two smaller surveying altimeters, and the fourth was a Kollsman aircraft altimeter, which would read to the nearest twenty feet. The Kollsman had been calibrated for internal friction one week prior to the expedition, and the error factor chart was taped to its side. Brucker and McClure drove down the gravel road leading from the cave, turned into the road to Mammoth Cave, then checked the topographic map of the nearest U. S. Geological Survey bench mark. The map showed one at the side of the next crossroad several hundred yards to the west.

They parked the car and got out carrying the instruments. After several minutes of searching in high weeds they found the bench mark, a low galvanized pipe filled with concrete, with a bronze disk imbedded in the top. The elevation was neatly stamped on it, 854 feet above mean sea level. One by one they set the altimeters to read 854 feet.

In five minutes they were back in camp; they had traveled fast to minimize the effect of changing air pressure due to weather. They established that the elevation of the desk top in the communications tent was 750 feet. This would be the base point for all readings, and later, when they were ready to take an altimeter beyond the Trap, they would leave the large altimeter at this point so they could check its reading at any given time by telephone. In this way, readings taken below could be corrected for existing pressure changes at the time of the reading. For some reason the two smaller surveying altimeters failed to respond at all to the change in elevation, so Brucker and McClure abandoned those and took the Kollsman down into the cave.

They took the first reading on the ceiling of the entrance, the flat under-surface of the Cypress Sandstone which was near the cave's roof rock. The altimeter read 682 feet. The next reading was at the bottom of Grand Canyon Avenue near the tomb of Floyd Collins. This was 618 feet. The two Rogers moved to the Valley of Decision, taking a reading at the bottom: 588 feet was the elevation. Their reading at Scotchman's Trap was 609 feet. They were aware that readings taken beyond the entrance would be influenced by any pressure differential which might exist between the cave atmosphere and the outside, but they felt that their readings were accurate enough to give some indication of the depth of the cave.

Brucker and McClure returned to the surface for lunch, then went to talk with Ed Comes, who looked up from his drawing board when they entered his tent. He told them that a preliminary plotting of B Trail and Bogardus Waterfall Trail indicated that they ran directly under the commercial route from Grand Canyon Avenue to Devil's Kitchen. That was startling news, since it was the first indication we had which pinpointed anything in the lower levels with passages above. Had Ed known this for long? Not very long. Had anyone been down looking over the commercial route in the light of this new information? No, it was too new. Did Ed have any ideas? Well, there was a shallow pit that he had his eye on in the right-hand passage at

The altimeters are set at the nearby bench mark. *Photo by John Spence.*

Cross Cave, just beyond the turn at Devil's Kitchen. Water'ran like a faucet from the top of a small dome and fell into the pit, eight feet deep, very much like the dome-pit spring on the tourist route a few feet away. The bottom of the pit was a basin, similar to the one at Ebb and Flow Falls. McClure and Brucker thought of the same thing as they turned to each other simultaneously. Is this the water for Bogardus Waterfall below? They were determined to find out.

They looked up Hugh Stout, the assistant geologist, and told him of their plan. They intended to run out to Bogardus Waterfall after they delivered supplies to Camp One. McClure would phone Stout from Camp One just before they started. At an appointed time Stout was to dump packets of fluorescein dye into the water in the basin at regular intervals. If the two Rogers saw vivid green water falling from Bogardus Waterfall, they would know there was a connection. Stout agreed to handle the arrangements from topside.

About 3:00 P.M. a party entered the cave with one Gurnee can for Camp Two. Brucker had the Kollsman altimeter carefully wrapped in a sweatshirt in his rucksack. Roger McClure and Lou Lutz led the way as Hugh Stout and Audrey Blakesley brought up the rear. They stopped at Cross Cave to investigate the pit at the end of the right-hand passage, a tunnel-like lead about four feet high throughout its fifty-foot length. Near the end, the sound of falling water became quite loud, so they crawled toward it and looked down into the pit. The bottom was shaped like a large bathtub of bright-gray limestone. Water splattered into it, then ran out through an eight-inch hole in the bottom.

They chimneyed down into the pit for a closer look. Pebbles rolled back and forth in the basin. Now and then one would be washed into the outlet. They tried to peer down this hole, but stones were wedged in, blocking their vision. The far end of the pit was a wall of breakdown of the same kind that sealed off Devil's Kitchen against further travel. If this water did form Bogardus Waterfall below, what a rascal Nature would be. It would mean that water would reach a point in five or ten minutes that man could reach only by eight or ten hours of hard work.

McClure, Brucker and Lou Lutz said goodbye to Audrey and Hugh at the Trap and started in. They took their first altimeter reading at a place on the supply route about a hundred feet beyond Ebb and Flow Falls, where a phone had conveniently been spliced into the line. McClure called Base Camp:

"Hello . . . McClure at Ebb and Flow Falls . . . no, everything's all right. Do you see the altimeter there on the desk? . . . Good. What does it read? . . . The outside dial . . . the one that's marked feet. . . . O.K. 745. You're sure of that? . . . Looks like you're in for some bad weather up there. Thanks very much. . . . Yes, one Gurnee can to Camp Two . . . O.K. . . . So long." McClure subtracted the present surface reading from the original reading to find the difference in feet to apply to the final reading. Brucker unrolled the altimeter carefully from the sweatshirt and handed it to McClure who looked at the dial, then tapped it gently with his finger. He read it again, then referred to the correction factors on the side under the column labeled "after tapping." He corrected the reading for mechanical error, then applied the differential from the two surface readings. Since they had no way of determining pressure lag, they had to ignore this factor.

"Six hundred is as close as I can come to it," said McClure. Brucker marked it in his notebook.

The trail to Camp Two was well marked by signs and telephone wire. It lay along the C Trail. After moving along slowly for some time they came abruptly upon several men. Bernard Wilt shouted a cheery greeting.

"Welcome to Camp Two!" he said. He was rolling up a sleeping bag. The Rogers stood in a narrow trench-like canyon looking to their left into a three-foot-high chamber about forty feet across. The floor was covered with red sand of the Flat Room variety, they observed, as they climbed out of the trench and crawled over to Wilt.

"Is this all there is of Camp Two? All I see is a couple of sleeping bags," said McClure.

"No," answered Bernie, "the main camp is down in the canyon back that way." He was pointing toward the back end of the room. "This is just the dormitory." At his invitation the three-man supply party crawled after him over the sand floor to the brink of the canyon. The telephone wire plunged into the canyon and an arrow on a rock pointed almost straight down. Brucker thrust his head over the edge and saw far below the warm glow of several carbide lamps. People were walking around in a small room about forty feet below.

"Looks great, but how do you get down?" Brucker asked.

"Follow me," said Wilt. He placed his hands on both sides of the canyon and began to chimney down using handholds and footholds that were not visible from the top. When they actually started in it was quite easy going down, they discovered, in spite of the open space below. In a few minutes they walked into a well-appointed Camp Two, complete with coffee boiling on a gasoline stove and a pantry full of food supplies. A swift stream bisected the room on its long axis and served as a source for drinking water. Phil Smith, Bob Handley, and Earl Thierry were already there. Soon Roy Charlton and the two marines arrived.

"We just received permission to close Camp Two," said Handley turning from the telephone.

"Then you won't need the supplies we brought," said Lou.

"No, I guess you'd better take them on to Camp One. If you wait a few minutes you'll have company going over. We're just about to close out here."

"We'll wait. Say, where's this big waterfall everyone's been raving about?"

"Up that way, but don't muddy the drinking water. We think it supplies the well for Bill Austin's house." Handley pointed upstream. "That's the way to the bottom of it. If you want to see the top, you go back up to the 'twentieth floor' and continue the way you were going." Lou and Brucker straddled the stream. Around the first bend they encountered a deafening roar of falling water, and after rounding two more bends they stood at the bottom of the fall. Water thundered down from above, splattering against rocks on the way. It was a regular downpour. Their lights couldn't penetrate beyond the falling water, and further reconnaissance could only be pursued at the expense of getting soaked. Brucker motioned to Lou that they ought to go up, for talking was impossible.

Cave acoustics play strange tricks. At the top of the canyon they were unable to hear the water until they had crawled around a bend in the passage, then the full sound of the water, an underground Niagara, assailed their ears. They crept to the brink of the chasm. From a high passage a torrent of water poured forth, like a fire hydrant opened wide. The water hurtled into the pit, striking ledges and rocks on the pit wall, and sending up a cloud of mist. For almost five minutes they lay bewitched by the display. Then it was time to go.

Lou and Brucker each carried a sleeping bag. Roger McClure had the Gurnee can. The best way to move the sleeping bags in the flat sandy room, they found, was to throw them ahead, then crawl up to them. They approached the canyon they were to enter, where a slab of rock with an arrow pointing into the canyon had a skull and cross bones marked on it. Brucker held onto the sleeping bag and crawled over to look in, recognizing the hole as being the top of X Pit, the deepest known pit in the cave. The warning was well placed.

"Is that where we go down?" yelled Lou from behind Brucker.

"No, we go down off to my right, there's a big pit here—" A sleeping bag plummeted past Brucker's left ear, hit with a smack on the edge of the slope leading into the pit, then rolled down into the blackness. Later a faint "thump" floated upward. Brucker looked back at Lou. Slowly he inched his way over to the canyon and peered in with a look of sheepish amazement.

"There's one sleeping bag that won't have to go to Camp One," Brucker remarked dryly. He crawled on, leaving Lou gazing into the pit.

At the end of the expedition Jacque Austin, Bill's wife, heard about the unfortunate sleeping bag affair. However, she was not one to deprive the society of revenue, and promptly sold Luthcr Miller a sleeping bag, sight unseen. After Miller had anteed up the coins, she told him he could claim his prize F.O.B. X Pit. The intrepid cave explorer is still wondering how he let a woman outwit him.

Bernie Wilt led McClure, Lutz, and Brucker over a new route he and some of the others had discovered between Camps One and Two. It led through a series of tight squeezes, gypsum crawls, flint crawls, chimneys, drop-offs, and other obstacles. Its total distance may have been shorter, but the time needed to traverse it seemed twice as long as the old way. About 8:45 P.M. they climbed out of Mud Avenue and signed into the Camp One log.

Luther Miller was singing into a telephone a song he had composed during the week, and the group at camp crowded around him, emitting loud guffaws when his lyrics covered an experience common to everyone in the party. He peeled off verse after verse, and everyone wondered how he remembered them all. He sang to the tune of the English ballad, "Bless 'Em All."

Late in the week Ida and Nancy's place resembles a hobo camp, but it never fails to provide tired support teams with steaming coffee. *Photo by Glen Shoptaugh.*

LUTHER'S SONG

There's a caving trip leaving today,
Bound for the dark underworld,
Mostly made up of young spelie recruits,
With trousers most clean and a shine on their boots.
They think they will show us some cave,
And I really hope that they do,
But all you can see is their hamburger knee,
While crawling down Flint Avenue.

Bless 'em all, bless 'em all,
The long, the short and the tall.
Bless Nancy Rogers when she's in Camp One,
You know that the beans are extremely well done.
Now when you reach camp you are tired,
You know that you've caved long and hard,
And if she weren't there, you would eat only air,
So cheer up, my boys, bless 'em all.

Bless 'em all, bless 'em all,
The long, the short and the tall,
Bless our dear Doctor and all of his pills,
There's one that will kill every one of our ills.
But when he puts pills in the drink,
It simply don't taste like you think.
I'm sure that it's fine, for it tastes like iodine,
So cheer up, my boys, bless 'em all.

Bless 'em all, bless 'em all,
The long, the short and the tall.
Bless our dear Ida when she is in camp.
When she goes exploring, directions get damp
And she roams around by the hour
Admiring each cute gypsum flower,
But when she is done, she is back in Camp One,
So cheer up, my boys, bless 'em all.

Bless 'em all, bless 'em all,
The long, the short and the tall,
Bless our Bob Halmi of *True* magazine,

His three thousand pictures will never be seen.
There's a flash in the passage ahead
And a voice that would waken the dead,
A "Son-of-a-gun, but this picture is rich,"
So cheer up, my boys, bless 'em all.

Bless 'em all, bless 'em all,
The long, the short and the tall,
Bless old "Pop" Dyer, he is slow and he's bald,
But I like going slow down a fifty-foot fall.
So let's say goodbye to them all,
As down the dark passage they crawl.
They'll soon quit their bragging, their tails will be dragging,
So cheer up, my boys, bless 'em all.

Luther Miller writes his song by the light of a gasoline lantern while Charlton looks on in amazement. *Photo by R. J. Richter.*

After supper Rog McClure phoned Base Camp for the latest altimeter reading which he used to determine the elevation of Camp One. The reading was 534 feet after applying all the correction factors. This would place the elevation of the Lost Passage about 230 feet below the level of Base Camp. This figure did not agree with that determined by Bill Austin's calculations on his return from the Mud Avenue exploration. By deduction he reasoned that the Lost Passage was situated at an elevation of 490 feet above sea level, or a difference from their reading of forty-four feet. What could explain the seeming error? First, we were all aware that in large cave systems there are lags between the change in air pressure outside and the change inside. Dr. E. R. Pohl, who had done extensive mapping in Mammoth Cave, had determined lags varying between three and thirty-six hours in certain passages. Without simultaneous barograph recordings on the surface and in the Lost Passage, we were unable to tell what the lag for this given situation might be. The meteorologist's smashed barograph couldn't help us.

Members of the society had made a number of air-pressure studies in smaller caves, but none in a cave system before. These pressure lags would be a problem for another expedition to tackle.

But Austin's figure was suspect too. A photo flashbulb lodged on a ledge in Mud Avenue would not necessarily establish the Green River level, since the bulb might have been left there by *receding* water instead of the crest. This would tend to place the elevation of the Lost Passage even lower than Austin's estimate. On the other hand, it was conceivable that the orifice connecting Mud Avenue to an underground tributary of the Green River might be small, resulting in local ponding in Mud Avenue above the Green River flood crest. In this event, the actual elevation would be much nearer that determined by the altimeter. Positive correlation will be the job of some future expedition to Crystal Cave.

McClure called Base Camp and asked to speak to Hugh Stout. The switchboard operator at Base Camp was unable to locate him for almost half an hour, but finally he came in. McClure estimated that by traveling at top speed, carrying only the altimeter, he could reach Bogardus Waterfall by 11:00 P.M. It was then 9:00 P.M. Stout would dump one packet of dye in the water above at 11:00 P.M. and another at 11:30 P.M. He promised McClure he wouldn't fail.

At five minutes to eleven Roger McClure and Roger Brucker chimneyed down water-covered rock walls to a muddy chamber. They followed a small stream a short distance, and sat down at the edge of a pool to wait. The waterfall, about twenty feet high, spattered on projecting rocks on its downward course and gave them a continuous shower bath of icy water. As much as they hated to stay in these cramped quarters, they had no choice in the matter, since from no other place could they see the pool or the falls. After going to all this trouble they had no intention of leaving, no matter how uncomfortable it was. At precisely 11:00 P.M. they unleashed a chorus of yells and loud whistles, hoping that if there were a connection, they might be heard by Hugh Stout somewhere above. If the timetable had been followed, he would be down at bottom of the basin pouring in the dye. Even if he heard, he might not be able to call back loud enough for his voice to carry over the din of the water.

Time: 11:05 P.M., no dye. The two Rogers opened a can of rations and settled back. Brucker had put them in his pocket just before leaving Camp One. 11:30 P.M.: they yelled again. Still no dye. They continued to wait, getting soaked more completely by the minute. Because they had moved rapidly through the passage coming out to the falls, they had perspired freely, and now were beginning to chill. They stood up and flailed their arms in violent movements to shake off the cold and dampness. Still no dye.

"Let's wait until 12:30, then go back, dye or no dye," said McClure. Brucker agreed, watching the minute hand creep around to their endurance limit.

At 10:50 P.M. Hugh Stout and Audrey Blakesley passed the spring at Devil's Kitchen and started into the Gypsum Route. At a place known as Cross Cave a hundred feet farther on, they turned into a right-hand passage. They dropped to their hands and knees and picked their way over breakdown blocks on the floor.

Stout slowly lowered himself into the basin. Audrey handed him two bright yellow packets of fluorescein dye. Stout ripped off the top of one packet and carefully laid it on a ledge for the time being. He checked his watch.

"Five of eleven. They ought to be there now. Maybe we ought to yell and make a lot of noise. If they're down there and it's not too far away, they ought to hear us," said Stout. They yelled at the top of their lungs, then listened. They heard no sound except the water splashing into the gray limestone basin and gurgling down through the crack in the bottom.

"It's time," said Stout. He poured the pink powder into the water, watching how it instantly dyed the water a blood red, changing slowly to a vivid fluorescent green. The entire pool was colored.

"Wow!" said Audrey. "I never saw such a bright green in my life! If they don't see that down below they ought to have their eyes examined."

"Brucker and McClure are probably seeing it about now," said Stout, "but we'd better put the other batch in at eleven-thirty just to make sure." The colorful green fluid continued to pour down the orifice.

Stout dumped the next packet in. The dye was stuck together in a cake and didn't dissolve immediately, so he plunged his hand in the water to break it up. Audrey climbed down to help. When they finished working in the emerald pool, they stared in surprise at their green- and crimson-streaked hands. In climbing out of the basin they managed to distribute the green hue liberally over their faces and clothing. When all the dyed water had gone down the drain, they hurried back to the communications tent to await word from below. Curious visitors came to examine the colorful pair.

On the dot of 12:30 A.M. McClure and Brucker painfully chimneyed out of the bottom of Bogardus Waterfall. Every muscle ached from sitting in such cramped positions. Their tiredness was matched only by their disappointment at not seeing any dye. So that the trip would not prove entirely unfruitful they read the altimeter at the bottom of the Bogardus Formation. Into the record went 526 feet. They took one more reading at the top of the Bogardus Formation at the end of the Lehrberger Link: 581 feet. At least they knew one thing: explorers had been chimneying up a fifty-five-foot formation all week and had been calling it a sixty-foot formation!

At Camp One a message waited for them to call Hugh and Audrey immediately on their return. Had they seen the dye? No, it hadn't come through by 12:30 P.M. The disappointment on the other end of the line was equally as great as theirs.

Fluorescein-dye testing to trace cave streams is a French innovation and has been used relatively few times in America. Several years ago Earl Thierry and I had used fluorescein successfully to study a complex stream system in Starnes Cave in southwest Virginia. Naturally, we were eager to try it on streams in Crystal. The dye is an organic compound, harmless to humans and animals, but its chief virtue is that a little goes a long way. One part of fluorescein in 100 million parts of water can be detected wherever it reappears again, and it is not absorbed by cave fills or by the calcium carbonate rock of cave walls. It will dissolve in both slightly acid and slightly alkaline water, so it is well suited for use in caves.

A study of European literature on speleology gave us some possible clues as to why we had not seen the dye. In England, the Yorkshire Geological Society dumped one and one half pounds of fluorescein dye in a small stream in a cave known as *Long Kin East*. Thirteen days later the dye reappeared in a spring only one mile away. After several unsuccessful attempts to trace the river in Padirac, in France, speleologists added 165 pounds of dye to the water. That was on a bright 22nd day of July. On the 24th of November of the same year the dye reappeared in a spring six and one half miles away. Short distance tests in Britain have shown that sometimes fluorescein dye takes up to three hours to travel four hundred feet, even under optimum conditions.

Plainly we had not waited long enough to prove anything conclusive. The invisible endurance barrier had won again.

19

DAYLIGHT

T HE CAVE WAS BEGINNING TO MAKE SENSE, at least to a few of the explorers. Ed Comes, the man assigned to plot the survey, was amazed
to see that horizontal distance in a straight line from Base Camp to Camp
One was only about 1,000 feet in spite of the fact that an explorer had to
traverse more than two miles of tortuous passages to get there. Camp One
lay about 230 feet below the surface of the earth from Base Camp. The
north end of the Lost Passage ended directly under the high bluffs flanking
the Green River. No wonder further progress was blocked by flowstone-
covered breakdown! How far was it through the breakdown to the outside?
It couldn't have been far, for Charlton had felt the rush of cold air coming
in.

The south end of the Lost Passage fell into position within a hundred
feet of a sinkhole in the surface. The presence of the sinkhole explained the
muddy breakdown that blocked the passage here. But Austin and Lehrberger had found smaller passages lower down threading through the rocks,
and perhaps another expedition could reach more cave beyond the sinkhole.

C-3 Waterfall, on the other hand, wasn't easy to explain. A topographic
map revealed no surface sinkholes anywhere near it, and the explorers
could only surmise that this was one of the tricks played by the Cypress
Sandstone. Water could enter this permeable sandstone cap rock and move
through a network of subterranean passages in the limestone until it
plunged over the falls. This would mean that the cave solution process was
still going on at a higher level somewhere to the south. Perhaps in the year
3000 a party of cave explorers will enter this new system and wonder how
it has been overlooked all these years. They may not dream that we sus-

pected its existence but held back because we didn't care to "get wet." Then, it is always good to leave something for others to do.

One phenomenon we had not counted on hearing from again was Earl Thierry's smoke screen. We were mistaken. A guide, while leading a tourist party over the Helictite Route on the upper levels, smelled smoke in the passage at the same time those below were coughing and rubbing their eyes. An investigation for a possible connection between the lower levels and the upper levels in the vicinity of the Helictite Route might profitably be made in the future, using controlled-smoke experiments.

Although the map had not been plotted all the way to Bogardus Waterfall, Austin smiled to himself, for he could see indications that might explain why McClure and Brucker had found no green water the night before. He guessed that Bogardus Waterfall lay farther south than we had previously suspected. His experienced eye told him that the final map would prove it.

Evidence was also piling up that one of the old pet theories would no longer hold. For years anyone acquainted with Crystal Cave had thought that the lower levels had been dissolved by the same water which had dissolved the limestone along the tourist trails. Now it was beginning to appear that some of the lower level passages had very little to do with the upper levels in terms of geological history. The lower level passages, particularly the Flat Room and the Lost Passage, appeared to be independent systems, apparently formed by water moving from a larger source to the south and west, from the heart of Flint Ridge.

Five areas in particular held the greatest potential for new discovery. The Bogardus Waterfall area, though heavily explored during the week, still had at least six unexplored leads on four levels heading into the ridge. Lehrberger, Charlton, and Miller had been the spearheads in that part of the cave. Early in the week they had given it up as being fully explored, but the news brought back by other parties had driven them back. Jack Lehrberger best expressed the situation when he said, "It's like a sponge out there. It just goes and goes."

The Bogardus leads pointed toward Salts Cave, which was itself a part of the extensive system. Somewhere out there, Crystal Cave explorers may find footprints of earlier men who penetrated to the northernmost reaches of Salts Cave. On that day, there will be more than a ripple of excitement in the speleological world.

A second area of excellent potential for future exploring is the Lost Paradise section where expedition members just tasted the thrill of discovery. Only two parties penetrated it, the first by accident and the second to describe it. No one tried to explore it. Passages lead out of this area in a southerly direction, again into the heart of Flint Ridge, and again to the endurance barrier.

An area not even examined by expedition personnel remains a mystery. Shortly after entering Scotchman's Trap the route drops down into a canyon. To the right lies the well-traveled route to Camp One and the lower levels. To the left, the canyon snakes off into darkness. Several months before the expedition Austin and Lehrberger strapped on knee-pads and fired up their carbide lamps to explore the canyon. They traveled for eight hours, crawling most of the time, in a direction they believed to be south. Out of food and energy, they stopped to look longingly at the continuing passage, then reluctantly retraced their steps. They reported finding passages leading downward, any one of which might open into a system comparable to the lower levels already known. But they agreed to follow the main passage rather than explore side leads. Perhaps fatigue and frustration are the price of persistence.

A fourth area to tempt the adventurous is down Mud Avenue to the source of the great underground roar heard by Austin. Ropes and dry clothes will have to be taken on this trip, and also a rubber boat. The idea of an underground river captures the imagination somehow; at least it did that of the press representatives. They kept phoning to quiz Phil Harsham on the progress in the search for the river, and he kept phoning back that all he knew was what he had been told. They grew impatient when reports of progress along this line were not forthcoming. One conversation between a reporter in Base Camp and Phil Harsham went like this:

REPORTER: How about this river you are looking for?
HARSHAM: There is no further exploration on it.
REPORTER: Looks like you are going to come up without it, huh?
HARSHAM: It looks like it. That would be a real find in here.
REPORTER: Are they going to look for it at all today?
HARSHAM: There are no plans for it yet. Joe Lawrence is down here. I don't know what his plans are. He may go looking for it. He is still sleeping.

But as expedition leader I had made the decision at the start of the expedition. This was a reconnaissance in force, not a major assault on one particular area of Crystal Cave. All the effort was not to be thrown in one direction at the expense of the others. There would be time for that on future trips.

The fifth area does not hold the promise of the others since it was fairly well explored by the personnel of Camp Two. Handley's Canyon has side leads, all unexplored, going toward an area of the ridge where there is no known cave passage. The reason it does not hold the potential is that there are not as many possibilities for exploration here. This is a judgment of the explorers and not necessarily a prediction, since it takes only one small crawlway to lead to big cave passages.

Three major "finds" had been made during the week. First was the Lehrberger Link cutting the time to Bogardus Waterfall from Camp One by two hours. The second was Lost Paradise with its crinoid fossil walls and spider-web complex of unexplored passages. The third was Handley's Canyon. Many side passages that had never been entered before were explored, and short cuts from one place to another were found. One of the biggest jobs accomplished was the collection of scientific data and the acquainting of sixty-four cave explorers with the Crystal Cave system and new techniques of exploration. Future expeditions will have a nucleus of trained explorers to call upon.

Two weeks prior to the expedition, Jack Lehrberger, while looking out a small crawlway near Devil's Kitchen on the commercial route, had discovered some Indian reed torches among the rocks on the floor. On examination, they were found to be about nine inches long and about one-half inch in diameter. Their tips were charred where they had been lighted. He discovered about half a dozen of these, and, though not an archeologist, he allowed them to remain in the position in which he had found them.

Since mud and water were present in this breakdown-littered passage, he believed that the torches had been discarded by Indians who had come through an entrance now blocked. Certainly they had not entered by the sinkhole Floyd Collins discovered, for Floyd had to dig his way into the cave. This belief was strengthened by the fact that reed torches had been found nowhere else in the cave.

Indian relics, including reed torches, have been discovered in other caves in the area. Both Mammoth Cave and Salts Cave have yielded archeological evidences of Indian visitation. We agreed, in the light of our findings during the week, that Indians probably had not penetrated beyond one isolated crawlway in Crystal Cave; if they had, they had left no traces.

In the glow of the lantern at Camp One we discussed these things while we waited for the hour hand to tick around to our scheduled departure time.

"I hate to leave," said Lehrberger. "Gonna miss Ida's cooking."

"I bet you will!" she said from her place near the stove.

Luther walked over to her. "Ida, I'm going to tell you something . . . something I don't tell to all the girls. You make the most delicious raisin oatmeal I ever ate in a cave."

Thirteen people remained in the Lost Passage. Most of them were finishing a hearty breakfast of oatmeal and chili, washing it down with steaming coffee. Skeets Miller had his notebook unfolded and was writing notes for his final broadcast from Crystal Cave. Harsham sat near him making notes for his coming conversation with reporters on the surface.

Barr and Comes examine charred torches left in the cave by Indians who had used them for light. This is the first evidence found that Indians had entered Crystal Cave. *Photo by John Spence.*

At eight o'clock I moved to the telephone with a cup of coffee and turned the crank. "Ray, this is Lawrence. Put me on tape. I have a report to record."

"Hold on . . . OK, go ahead, Joe."

I reported as follows:

At 8:55 A.M. on February 20th, operations in the cave are complete except for the withdrawal of personnel and valuable equipment. Yesterday the exploration mission of Camp Two was completed. Group leader Bob Handley recommended that Camp Two be closed. Orders were issued on this recommendation and

Camp Two was abandoned. Personnel were transferred to Camp One and to Base Camp. We are now in a position to summarize some of the activities of the expedition. Exploration has been concentrated in these main areas. The Bogardus Waterfall-B Trail was worked first. The leads above Floyd's Lost Passage were explored. A number of leads in this area were not completely looked into. The area around C-3 Waterfall was explored from Camp Two. The explorations yielded a lot of cave which hadn't previously been entered by men. Discoveries included pits, domes, waterfalls, and canyons. One of the largest canyons discovered on this expedition was found by Bob Handley in the Camp Two area. This canyon was close to a hundred feet deep. A special summary on the scientific efforts of the expedition has been issued by the scientific staff members.

Surveying was undertaken in the main areas of exploration. The surveyors ran a traverse through the explored areas forming a backbone which makes it possible to prepare a reasonably accurate map of the cave. Many of the instruments used in the survey were loaned to the expedition by the Ohio Division of Geological Survey.

One of the purposes of this expedition was to learn how an expedition of this size could be conducted. We have gained many valuable lessons from our successes and our failures.

One of our failures was the lack of full coordination between surveying and exploration. We found that it is very necessary for survey notes to be plotted immediately and made available to explorers so that they can use that knowledge to push forward. Our machinery for doing this was not always in operation.

One of our outstanding successes in this expedition was the telephone system which our communications officer, Bob Lutz, designed and set up.

An error in command cost the expedition the services of one party for about two days. This occurred due to lack of knowledge of the physical effects of a long stay underground on the individual. As has been mentioned in an earlier report, the varied sleeping and eating schedule of the individuals underground caused a loss of energy due to lack of sleep and insufficient food. Three days after the expedition started, it became evident to the leaders of the expedition that practically the entire underground crew was suffering from this handicap. To safeguard the health of the expedition, some members were ordered out of the cave to recover. Two days later, at a time when the expedition leader returned underground, the expedition members had adjusted themselves to the unfamiliar eating and sleeping conditions and began to recuperate. Those who had lost considerable energy recovered almost completely and were able to continue in their activities underground.

About the middle of the week, when morale seemed to drag, Group Leader Dyer began to inject humor into the Camp One life. This was a definite boost to morale.

We have learned much about logistics and supplies on this expedition. One of the outstanding equipment developments was the Gurnee can which was designed by Russell Gurnee and John Spence. These canisters made of sheet metal and pointed at the nose can be dragged through the crawlways quite easily. The canisters are oval in cross section, being about 10 inches on the long diameter and 8 inches on the small diameter. The total length is 24 inches. Of all methods tried to bring in supplies, we found the Gurnee cans far superior to any other containers used. This is a notable development in subterranean logistics.

I put the phone back in its case and looked at the inactive group. Nancy Rogers was scraping the last of something out of a big pot. Brucker was staring off into space at nothing. His eyes were bloodshot and watery, a sure sign of fatigue. It was high time we were pulling out. George Welsh grinned as he rocked back and forth on his heels, then slowly lit a cigarette. I decided to send him out with the first party.

A lively debate started when Ida wanted to know what to do with all the food and utensils in the kitchen. Phil Smith said we should leave everything here since the Ohio explorers planned to come back to do some more rummaging around in the cave. The supplies would come in handy, he said. I gave orders for the dishes to be washed and all waste paper to be piled in one place up the passage for burning later, *after* people were away from the area. Everyone moved slowly, but by degrees the kitchen was placed in order and for the last time there were no more dishes to wash.

Explorers descended like pack rats on what remained of the candy bars and peanuts. Raisins went into inside pockets, already bulging with odds and ends of personal gear. I dispatched a party to our sleeping quarters to bring back to camp all sleeping bags and air mattresses. We planned to take the mattresses out with us and any of the sleeping bags in condition to stand the trip.

"Hey, look what I found!" Lou Lutz was coming back from the north end of the Lost Passage with a sleeping bag draped over his arms.

"I thought we had all of them," I said.

"This one looks like it's been used only once or twice. Found it back in the passage behind Fool's Dome."

Explorers had stopped sleeping back in the quiet cut-around tube early in the expedition. It turned out to be too remote from Camp One for hungry explorers on arising. This bag had evidently been left there and hadn't been used since. How many weary cavers had been deprived of a good night's sleep?

Skeets came over to me and requested permission to leave the cave in the last party so he could make a final broadcast describing Stygian darkness taking over the camp, or something like that. I told him the press would go out with the same guides that had taken them in and that he would have the privilege of looking back on darkness.

Tired backs were bending over sleeping bags, wrapping them up into lumpy rolls, a far cry from the sleek compact bundles that had been brought in way back on February 14th. The "whoosh" of air from the mattresses being deflated was the only sound during this activity. Brucker went down the passage and came back with a length of wire which he bound around a bundle of four bags. He hoisted them to the ceiling, twisting the wire around a piton which had formerly been used to hold a telephone line off the ground. Now it didn't matter if the wire dragged on the floor. There weren't many people around to step on it. Wilt stuffed collapsed mattresses into one Gurnee can while Lehrberger tied the drawstring of another already loaded. Here and there bearded men picked through piles of personal equipment like so many misers counting out their wealth. They packed away knick-knacks, waterproof match containers, karabiners, and nylon ropes. Now and then they picked up an item and examined it closely, then they would withdraw something already packed, replacing it with the new treasure. I noted particularly that each man made his own neat pile of expendable supplies, usually on a shelf of rock where he could find it if he ever came back again.

At ten o'clock Luther called to Welsh and Lehrberger that they had better get started if they wanted to get out of the cave in time for lunch. Off they went, wagging their Gurnee cans behind them in zigzag trails on the sand. In ten minutes three more followed them. As they moved out in single file over the billowing mounds of sand, Halmi exclaimed, "Son-of-a-gun! I've got to get a picture of that," and he pranced off to a vantage point to photograph the departing column. He had taken over 1,200 black and white pictures and 400 color shots, but he was still at it.

Eight of us left. Is this how it feels in death row when the guards take prisoners away one by one? I thumped myself hard. Lawrence, I said to myself, you're getting cave happy.

I was about to leave with my party when someone discovered we were short one pair of knee-pads. A person could undoubtedly make it out of the cave without knee pads, but the result would be hamburger knees made so famous by Luther Miller's song. We spread out to scour the passage for missing equipment. Jim Dyer suggested that we flip a coin to see who would go out without knee-pads, and that the loser could very easily go through the crawlway on his back, doing the grasshopper walk. Thanks to Doc Wanger one of us was spared that pleasure. In his pile of surplus gear we found one battered pair of elbow pads. Since Nancy Rogers had the smallest knees, she was unanimously elected to wear them. She protested the rigged voting, but slipped them on like a trooper. As we pulled out she mumbled something about cut-off circulation, not being able to feel anything, and gangrene. Behind us five tiny dots of light marked the site of Camp One.

As our party passed around the bend in the passage, Brucker turned to survey what remained of the camp. With the gasoline lantern gone it was a dreary place with deep shadows. Bundles of sleeping bags swung crazily from the ceiling, turning gently in the slight breeze that had started from the north end. Phil Harsham was curiously looking over the pharmacy of pills and liquids Doc had left, reading the labels in the dim light. Dyer thumbed aimlessly through the log. Time dragged. Brucker looked into the

On the last day Nancy Rogers sorts through the equipment she will leave. *Photo by Robert Halmi.*

blackness, knowing that no explorers would come seeking food and company. He walked over and shoveled sand into the spent carbide pit.

Few words were spoken. Inside he felt sad that the end had come, and the others reflected emptiness in their faces. Jim Dyer had tears in his eyes, —or was it fatigue? Rationally, we would be glad to return to the comforts of civilization; showers, regular meals, restful sleep. But the strong bond of companionship would be hard to give up. We had worked and struggled together toward the same goal. We had laughed at our own discomforts. Skeets and Phil felt the *élan* that binds cave explorers together.

Roger Brucker felt something ought to go into the log, a few words that would be fitting for the final hour of a noble undertaking. In the log he wrote:

"And so we leave the cave in darkness, as we found it. We have unlocked many of nature's secrets, and still others are yet unsolved. Men will be back, probably to the end of time, but history will record the monumental effort of The National Speleological Society in the Year of Our Lord 1954."

We left the cave.

When Roy Charlton pauses to talk to Burton Faust just after coming to the surface, the effects of spending 156 continuous hours beneath the earth are evident in his face. *Photo by S. A. Loyd.*

EXPEDITION ROSTER

Name and Home	Occupation	Expedition Assignment
Austin, William T. Cave City, Ky.	Cave manager	Asst. expedition leader
Barr, Thomas C., Jr. Nashville, Tenn.	Zoologist	Biologist and geologist
Blakesley, Audrey E. Trenton, New Jersey	Stenographer	Rock climber
Bruce, C. N. New Castle, Pa.	Business executive	Photographer
Brucker, Roger W. Yellow Springs, Ohio	Scenario writer	Explorer
Charlton, Royce E., Jr. Dillwyn, Va.	Farmer	Explorer
Comes, Edward P. W. Caldwell, N. J.	Plant manager	Cartographer
Coulson, Rhoda M. Washington, D. C.	Business executive	Stenographer
Cournoyer, Donald N. Arlington, Va.	U. S. Weather Bureau employee	Hydrologist and meteorologist
Ellis, James G. Maceo, Kentucky	Engineer	Support
Douglas, Henry H. Falls Church, Va.	Cabinetmaker	Communications
Dyer, James W. Columbus, Ohio	Cave consultant	Group leader
Dymond, Louis T. Washington, D. C.	Museum guard	Communications
Feder, Dr. Ned Philadelphia, Pa.	Physician	Asst. physician
Faust, Burton S. Washington, D. C.	Patent examiner	Information officer
Gardiner, Ann H. Martinsburg, W. Va.	Director of nursing education	Nurse
Gehring, Dr. John R. Philadelphia, Pa.	Physician	Asst. physician
Gurnee, Russell H. Tenafly, N. J.	Sheet metal contractor	Explorer
Halmi, Robert New York, N. Y.	Photographer	Photographer for *True* magazine

Name and Home	*Occupation*	*Expedition Assignment*
Handley, Robert H. Charleston, W. Va.	Draftsman	Group leader
Harsham, Philip Louisville, Ky.	Newspaper reporter	Press pool reporter
Irwin, Ethel Keezletown, Va.	Attorney	Stenographer
Kacsur, Charles Staten Island, N. Y.	Schoolteacher	Support
Kendall, Lt. John P. Quantico, Va.	U. S. marine	Support
Klein, Marguerite M. Chevy Chase, Md.	Schoolteacher	Asst. meteorologist
Lawrence, Joseph D., Jr. Philadelphia, Pa.	Electronic engineer	Expedition leader
Lawson, William Louisville, Ky.	College student	Support
Lehrberger, John J., Jr. Louisville, Ky.	Dance instructor	Explorer
Longsworth, Lt. Chas. R. Quantico, Va.	U. S. marine	Support
Loyd, Mrs. Betty Waynesboro, Va.	Housewife	Asst. supply officer
Loyd, Samuel A., Jr. Waynesboro, Va.	Engineer	Explorer
Lucas, Ann Jane Martinsburg, W. Va.	Nurse	Nurse
Lutz, Robert L. Elkins, W. Va.	Electronic technician	Communication officer
Lutz, Ulysses E. Philadelphia, Pa.	Real estate broker	Rock climber
McClure, Roger E. Columbus, Ohio	College student	Explorer
Miller, Luther Oblong, Ill.	Farmer	Explorer
Miller, William B. New York, N. Y.	NBC executive	NBC reporter
Mueller, Albert C. Scotch Plains, N. J.	Supervisor	Explorer
Mueller, Mrs. Margaret Scotch Plains, N. J.	Housewife	Stenographer
Neidner, James H. Buechel, Kentucky	Administrative planner	Support

Name and Home	Occupation	Expedition Assignment
Perry, Kenneth M. Falls Church, Va.	Curator	Communications
Roudabush, Charles E. Washington, D. C.	Electronic supply	Storekeeper
Richter, Robert J. Brooklyn, N. Y.	Research chemist	Asst. biologist
Rogers, Nancy Arlington, Va.	Bacteriologist	Explorer
Sanders, Richard S. Wytheville, Va.	College student	Switchboard operator
Sawtelle, Ida V. Brooklyn, N. Y.	Dog trainer	Rock climber
Shoptaugh, Dr. A. Glenn Indianapolis, Ind.	Physician	Asst. physician
Smith, Philip M. Springfield, Ohio	College student	Explorer
Snell, Alden E. Arlington, Va.	Tax analyst	Supply officer
Spaulding, Carl Chevy Chase, Md.	Mathematics analyst	Support
Spence, John L. Brooklyn, N. Y.	Consulting engineer	Chief photographer
Stairs, Edith Somerset, Pa.	Bookkeeper	Support
Stephenson, William J. Bethesda, Md.	Patent examiner	Asst. expedition leader
Stone, Dr. Ralph W. Harrisburg, Pa.	Geologist	Consulting geologist
Stout, George H. Cambridge, Mass.	Graduate student	Asst. geologist
Streib, Mrs. Barbara Valley Station, Ky.	Housewife	Stenographer
Streib, Raymond Valley Station, Ky.	U. S. soldier	Switchboard operator
Talis, Albertine Philadelphia, Pa.	Bibliographic researcher	Switchboard operator
Thierry, Earl M. Charleston, W. Va.	Civil engineer	Chief surveyor
Wanger, Dr. Halvard Shepherdstown, W. Va.	Physician	Expedition physician
Wanger, William H. Shepherdstown, W. Va.	College student	Support

Name and Home	*Occupation*	*Expedition Assignment*
Watkins, William H. Annandale, Va.	Government sales representative	Support
Welsh, George E. Louisville, Ky.	Mover	Support
Wilt, J. Bernard Washington, D. C.	Bricklayer	Support

by Joe Lawrence, Jr.

ORGANIZATION REPORT

THE ORGANIZATION OF AN EXPEDITION is based on its objectives, and initial planning is aimed at setting down in detail just what is to be accomplished. Then the time, personnel, and equipment necessary to achieve the objectives are estimated.

I. PURPOSE

The objectives of the Floyd Collins' Crystal Cave Expedition were to conduct a study of
1. The cave topography through exploration and mapping.
2. The cave biota.
3. The cave geology.
4. The cave meteorology and hydrology.
5. The physical effects on explorers of a long, strenuous stay underground.

II. ORGANIZATIONAL STRUCTURE

These objectives and the nature of the cave dictated the structure of the Floyd Collins' Crystal Cave Expedition. Because of the comprehensive nature of the expedition objectives, about 25 people were needed to work directly on the achievement of these ends. Over half of this manpower explored and mapped the cave. Two or three persons were assigned to the achievement of each of the other objectives. Most of these 25 persons working directly on the study of the cave had to be maintained far underground. This posed a logistical problem that could be solved only by a large sup-

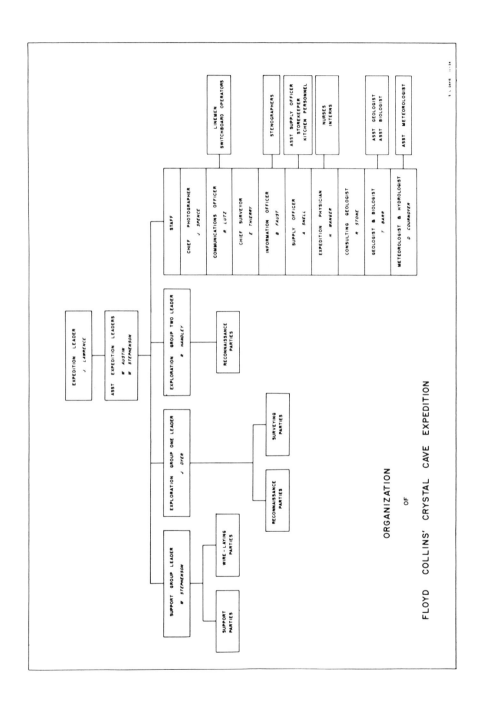

ORGANIZATION

OF

FLOYD COLLINS' CRYSTAL CAVE EXPEDITION

port group. Thus, the personnel requirements of the expedition exceeded 60 people.

The study of a cave can be divided into three phases—planning, field work, and evaluation. Planning proceeds from the statement of purpose through the selection of equipment and personnel and the outlining of field work to be done. Field work involves the collection of data and specimens in the cave. The third phase, evaluation, overlaps field work in time. Preliminary evaluation begins immediately after the first data are secured because additional data-taking may be influenced by preliminary results. However, the main job of evaluation is done after the expedition has withdrawn from the cave when data and specimens are carefully examined, final maps are prepared and reports are written.

The leader must combine the elements of time, personnel, and equipment through all three phases of the study to achieve the purposes of the expedition. Success on a large scale operation such as the Floyd Collins Crystal Cave Expedition requires the delegation of responsibility and authority to specialists. Each specialist is made responsible for a specific activity, such as leading, studying the biota, or procuring supplies.

Responsibility was divided between two groups of officers—command officers and staff officers. Command officers were responsible and had authority for directing the main operations. The staff officers were responsible for specific activities requiring special knowledge or skill and served as advisers to the command officers.

There were three levels of command: expedition, group, and party. Control of the expedition rested with the expedition leader and two assistant expedition leaders. Since the expedition operated simultaneously at several locations on different projects, it was necessary to create a subdivision called a group. A support group based at the cave entrance was to deliver food and equipment to those working in the cave. Groups at both Camp One and Camp Two, each under the direction of a group leader, were responsible for the study of the cave in their areas.

Specific missions in the cave were performed by parties under the direction of party leaders. It was the responsibility of the group leaders to organize parties, appoint party leaders, and assign missions. Experience has taught that a party of four explorers is the largest that normally functions efficiently. A party of two explorers is the smallest that can operate safely; under some conditions even it is too small to be safe.

Staff officers were responsible for scientific study of the cave and for the technical aspects of support. The biologist and geologist, the meteorologist and hydrologist, the consulting geologist, the chief surveyor, and the expedition physician were responsible for specific studies. The expedition physician, the chief photographer, the supply officer, the communications officer, and the information officer were responsible for support functions. The duties of the staff officers were itemized in detail.

The biologist was to:

1. Collect and prepare specimens of the cave biota.
2. Maintain the equipment necessary for proper collection.
3. Identify collections and prepare the biology section of the expedition's final report.

The geologist was to:

1. Make a surface study of the area.
2. Examine the structure of the cave.
3. Ascertain the various constituent stratigraphic members and their dip and strike.
4. If possible, collect or make molds of fossils in the cave.
5. Study the cave mineralogically.
6. Advise the expedition leader of unsafe areas of the cave.
7. Prepare the geology section of the expedition's final report.

The hydrologist and meteorologist was to:

1. Measure humidity, temperature, and air flow in the cave.
2. Study the cave water, ascertaining particularly the rate and direction of flow, temperature, chemical content, source, and destination.
3. Write the hydrology and meteorology section of the expedition's final report.

The consulting geologist was to:

1. Give technical advice to the expedition leaders and staff officers.
2. Review the expedition's final report for technical accuracy.

The chief surveyor was to:

1. Assist the leaders in coordinating the work of the survey parties so that as much of the cave as possible would be mapped.
2. Standardize surveying techniques.
3. Supervise preparation of the final map of the cave.

The expedition physician was to:

1. Examine all members for physical fitness.
2. Provide medical attention to expedition members.
3. Notify the expedition leader of all conditions that endangered the health of the expedition. To this end, the physician was to conduct such inspections of kitchen, water supply, and toilet facilities as he deemed necessary.
4. Prepare the expedition's final report on the physical and psychological aspects of the exploration.

The chief photographer was to:

1. Obtain a photographic record of the cave and the expedition.
2. Cooperate with scientific personnel in obtaining a photographic record of their findings.

The supply officer was to:

1. Procure, store, and issue expedition equipment and supplies.
2. Manage the expedition funds.
3. Administer the Base Camp mess hall.

The information officer was to:

1. Maintain a record of the expedition.
2. Be responsible for contacts with the press.

The communications officer was to:

1. Establish, operate, and maintain telephone communication between points designated by the expedition leader.

III. LESSONS LEARNED

The time lag between phoning survey notes to the surface and receiving maps in the cave was too great for the maps to be of use to the exploring parties. This was a great handicap to explorers since they did not know the true position of passages they had already explored. In future expeditions this difficulty may be overcome by plotting notes in the cave directly on enlarged topographic base maps.

Exploration is not complete until a survey has been made. The plan of operation was for a reconnaissance party to become familiar with an area of the cave on one day and for a survey party led by a member of the reconnaissance party to survey the area on the following day. This plan was followed in the Bogardus Waterfall-B Trail area. It was later found that reconnaissance and survey could be carried on simultaneously by the same party. This was done successfully in the Flat Room. In future expeditions, an effort should be made to reduce the ratio of reconnaissance time to survey time.

Rapid, detailed, and accurate communication of information from one individual to another, either orally or in writing, is essential for a prompt evaluation of findings and the coordination of efforts that should be based on these findings. The tools of communication available to the Floyd Collins Crystal Cave Expedition were an excellent telephone system, logs, written messages, and briefing meetings. The tools were adequate, but they were not utilized fully. The inadequate flow of information decreased the efficiency of the expedition. Poor coordination resulted from lack of maps, incomplete log entries, insufficient briefings, and a lack of familiarity with the known portions of the cave by many expedition members and a lack of common nomenclature for distinctive landmarks. The problem is much greater on a large expedition than on a smaller one. The fact that this was a first attempt at an underground expedition of such size explains why many, including the leaders, failed to appreciate the importance of a rapid and accurate exchange of information. As the ratio of expedition size to duration increases, adequate communication becomes more difficult.

The communication problem can be eased by a fuller use of the tools of communication, by keeping the ratio of expedition size to duration as small as is compatible with the objectives and resources, by briefing sessions, and by naming features in the cave and identifying them with signs or notes placed conspicuously near them. This last technique of leaving identifying notes in the cave was used by Robert H. Handley with notable success in the Camp Two area.

Only the very smallest caves can be studied adequately on a single visit. The National Speleological Society had learned that the study of a medium-sized cave required repeated visits. A small group would explore and record cave data on a week-end, finding obstacles they could not overcome and questions they could not answer. The evaluation of their findings would yield new questions and indicate possible solutions. Then additional trips would be necessary to advance the study of the cave. It was suspected that there would also be a need for repeated trips when studying large cave systems with large expeditions. The findings of the Floyd Collins Crystal Cave Expedition confirm this. A large expedition would have been impractical had it not been for the groundwork of the earlier explorations of the cave. While a large expedition produced a great deal of information about the cave, it also turned up some questions for future expeditions to answer.

Two of the outstanding successes of the Floyd Collins' Crystal Cave Expedition were in support. Both telephone communications and supply achieved results that many would have thought impossible. These two smoothly run activities owed their success to detailed advance planning, competent personnel, good administration, and the cheerful willingness with which the support parties expended their energy to conquer the Crawlway.

The information contained in the following Staff Reports is our legacy to speleologists who will assault large cave systems in years to come. We hope they will profit by our mistakes and experience the deep personal satisfaction we found in just one of the caves beyond.

2

by Halvard Wanger, M.D.

MEDICAL REPORT

Preparations and plans for the medical aspects of the Crystal Cave Expedition began several months prior to the scheduled date of entry into the cave. After a vast amount of reading the doctor became aware that little research had been done concerning studies of groups underground, the effect of prolonged humidity upon individuals in caves, and other phases of the physiology of speleologists. This type of research, therefore, presents a wide-open field to any qualified researcher.

I. PRELIMINARIES

The first problem which presented itself was that of screening expedition personnel. Only persons in good health were eligible. Eligibility was primarily determined by each applicant's own statement on the original application as to the condition of his health. To be considered qualified those accepted were required to submit summaries of their caving experience which were accepted as evidence of qualifications. Only one applicant was rejected because of physical condition, which seems to indicate that cavers as a group are above average in health.

Of course, one's initial acceptance did not necessarily guarantee that he would go underground. A rigid physical examination at the base of operation was the criterion that determined whether one would be chosen for underground work (unlimited personnel) or would remain topside (limited personnel) to perform the many duties which would keep the expedition going.

253

Pre-planning included recommendation for typhoid and paratyphoid inoculations. Tetanus immunization was made obligatory.

Several weeks before the trip a list of calisthenics was devised and recommended to all personnel in order to build up stamina and prevent sore aching muscles. One difficulty arose in planning such a set of exercises. Caving uses practically every muscle in the body, while almost all other activities tend to localize the main exertion to one area. Exercises strengthening leg, back, shoulder, abdomen, and arm muscles were stressed. One novel exercise, which involved crawling back and forth beneath a series of rather low chairs was given to simulate conditions found in most undeveloped caves. The caver was instructed to continue crawling and squirming until he was winded. Practical application of the "chair crawl" resulted in several comical situations, when some expedition personnel became wedged beneath the chairs and had to call for aid from amused friends and relatives.

Of the twenty-eight who followed the prescribed exercises, only nine experienced soreness and stiffness during or after the expedition; of these nine, four had followed the exercises only partially. Of the twenty who did not follow the exercises, ten noticed soreness and stiffness, whereas ten had no ill effects whatsoever. This seems to indicate that while the calisthenics were not absolutely foolproof, those who followed them rigidly fared better than those who did not. A suggested pre-expedition diet was offered to those who were overweight. Ten "unlimited" personnel lost an average of eight pounds each, and three "limited personnel" lost an average of eleven pounds each before the trip started.

Through conversation with Robert H. Flack, who had been in the cave as far as the site for Camp One, the expedition physician gained familiarity with the problems which might be encountered in the cave and with the type of medical supplies that might be needed. Camp One was to be the "jumping off place" of the exploring parties. From Mr. Flack's color slides the expedition physician gained an accurate impression of the type of passages to be traversed.

A tortuous 1300-foot crawlway presented a major problem in the event of an injury to a member of the expedition. Several theories were formulated, but it was never decided how a seriously injured person would be removed from the cave. Any method of removal would depend, of course, on the nature of the injury and the obstacles between the victim and the entrance. There were three ways of dealing with an incapacitating injury:

1. A shaft might be sunk from the surface directly to the victim.
2. A victim could be dragged out in a sleeping bag, protected by a plaster of Paris cast or shell.

3. A patient could be treated in the cave until he could come out under his own power. A fracture, however, would have kept the victim in the cave beyond the close of the expedition.

Fortunately, decisions of this sort were not required. There were no serious accidents.

It was felt the diet underground would be deficient in vitamins, minerals, and trace elements, so Roerig's contribution of their Viterra capsules was gratefully used as a dietary supplement.

As a result of viewing a motion picture entitled "Gateway to Health" (produced by the National Apple Institute) the physician decided that a timely contribution of ten bushels of apples from Appalachian Apple Service, Inc. would provide dental hygiene for explorers who could not conveniently use tooth brushes. Conclusions drawn from research work of Dr. Fred Miller, Altoona, Pennsylvania, demonstrated that a person may eat what he wishes but should cleanse his teeth afterwards with a fruit or vegetable; Miller's preference was an apple.

While we were well supplied and prepared to carry out this theory of dental hygiene, we realized that the weight of apples would prevent their being transported in large amounts to Camp One. They were stored inside the entrance of the cave, however, and were used extensively by the personnel topside. The stored apples presented a welcome sight to explorers on their way out and were eagerly eaten by those who craved fresh fruit.

II. CAMP MEDICAL ACTIVITIES

The expedition physician arrived at the cave several days prior to the major descent underground. This early arrival was necessary for three reasons:

1. An infirmary had to be established and medical supplies had to be separated into two groups—those to be sent underground and those to be kept on the surface.
2. The physician wished to make a preliminary trip to the site of Camp One in order to become personally familiar with the obstacles that the cavers would encounter.
3. Physical and psychological examinations of all personnel had to be given before the expedition went underground.

After arrival of the medical staff on Thursday, February 11th, the infirmary was established. Details were left to the nurses, while the physician made his "quick trip" in and out on Friday. That "quick trip" took nine hours. It was ironical that after the worry about insufficient medical personnel, two nurses were secured through the courtesy of the King's Daughters

Hospital, Martinsburg, West Virginia, and three internes (NSS members) from Philadelphia General Hospital were added to the staff. These were in addition to the expedition physician. Since the expedition was somewhat overstaffed, medically, the internes spent most of their time running supplies into Camp One, fighting forest fires, and doing K.P. There was always at least one interne topside, however, maintaining a twenty-four-hour infirmary in case of casualty in that area. The expedition physician stationed himself at Camp One, where light duties permitted his spending some time in stereo photography and exploration.

On Saturday the actual qualifying tests began—psychological, physical fitness, and physical exams. The physical examination, accomplished the day before the formal entry underground, included the filling out of psychological questionnaires as well as those pertaining to past medical history. Previous reading in the Armed Forces Medical Library and the Library of Congress had revealed no satisfactory psychological tests other than the Manifest Anxiety Test of Dr. M. J. Freeman, psychologist of North Hollywood, California. This test is described in a monograph, published by the American Psychological Association in 1953. Dr. Freeman kindly granted us permission to use his tests which included 141 questions. These questionnaires constitute the beginning of an extensive study of the psychology of caving. The medical history form included questions concerning past and present diseases, current symptoms and diseases, pertinent family history, past surgery and accident, and a statement as to present status of tetanus and typhoid immunization.

The physical examination performed was a routine one with the addition of three special features deemed appropriate for our purposes:

1. The vital capacity test was given. This is a measure, in cubic centimeters, of the quantity of air which a person can expel from the lungs by forced expiration after the deepest possible inspiration. This test is a valuable index of cardiopulmonary efficiency.

2. A variation of the Harvard Step Test was used. It involved stepping on and off a box of a size proportional to the individual's height. The examinee stepped on and off the box thirty times a minute for five minutes. Pulse rate was taken at three different intervals following the conclusion of the test. When these were added a score was determined which gave another valuable physical fitness index of the recuperative powers of the heart.

3. A petechiometer test, a test of capillary fragility, was used to give an index of a tendency to bruising. This test is performed simply by applying a suction cup apparatus to the skin of the forearm, using a

measured vacuum, and counting the number of petechiae (minute hemorrhages) within a circle one centimetre in diameter.

The expedition's medical supplies were carefully selected. A certain quota of gastro-intestinal upsets and upper respiratory infections were expected, so the usual medications for these conditions were procured. Wyeth Laboratories, Inc., gave the expedition six bottles of Plavolex, a dextrin plasma volume expander, which is a satisfactory substitute for blood plasma to be used in emergency treatment of shock. The same company provided the expedition with quantities of fungicidal preparations for instances of fungus disease which were anticipated due to the abnormally humid atmosphere in caves. Chas. Pfizer & Company, Inc., not only sent one of their biochemists to join our scientific staff, but also provided us with Terramycin and other antibiotics and purchased a substantial amount of medical equipment that the expedition had not been able to afford.

The Sisters of the Holy Ghost, who operate the King's Daughters Hospital, provided a variety of first-aid supplies. Plaster of Paris bandages were taken in quantity for possible fracture immobilization, although plastic is superior and is recommended for use in the future. A very simple apparatus for administering artificial respiration, supplied by the Mine Safety Appliances Company, was included among our supplies. This bellows-like arrangement with a face mask is very compact, being housed in a cubical eight-inch box. No manipulation of the patient's body is needed with this particular form of resuscitation, so it is of special value in confined spaces such as one finds in caves.

The actual medical administration began with the expedition's entry underground. Headaches, colds, gastric hyperacidity, intestinal upsets and exhaustion (one resulting in a mild case of asthma), numerous cases of insomnia, and many minor cuts and abrasions, some irritated by sand and silt, occurred. There were no serious mishaps, however; despite the elaborate medical preparations the three most-used medications were APC (an aspirin compound), tincture of benzoin, and mild sedatives.

Perhaps the most publicized casualty of the expedition was caused by fatigue. In one party of four men and two women to reach Camp One, three of the men and one woman experienced almost complete exhaustion. This was due to lack of previous sleep, slightly lower than average expeditionary health standards, and, most of all, an overload of supplies which the group was transporting to Camp One. This party also met several psychological blocks, one of which was the encounter at the Keyhole when an outgoing party informed them that: (1) there was starvation in Camp One due to the supply group fighting forest fires topside, (2) they were only one-fifth of the way in, and (3) that they would never find their way without a guide. Due to these and other circumstances, their trip was stretched into fourteen hours, a long time to be attempting to reach an objective, especially when there is exceptionally rugged territory the entire way.

The ensuing case of asthma, brought on by excessive fatigue and further aggravated by damp clothing and cold, was considerably milder than reported in the newspapers. The patient left the cave only because she was unable to meet the strenuous exertions which would have been required had she remained underground. The attack was not severe enough to require an injection of adrenalin, which was in readiness. The victim left the cave completely under her own power in only four hours, ten hours less than it had taken her to come in.

Water for drinking and for bathing was obtained underground from a spring several hundred feet away from the main section of Camp One. Bathing was accomplished downstream from the spring on a sandy beach.

The water supply at Base Camp on the surface was tested by the Kentucky State Department of Health and found to be free from contamination. The mess-hall attendants were under advisement from the local health officer, a sanitary engineer, as to proper operation of the topside facilities.

Seventy-six percent of personnel took the recommended typhoid shots. It was also urged that halazone be used in all the water that was drunk or obtained from underground sources to guard against the other diseases often prevalent in untreated water. Danger of water contamination was increased by the heavy traffic of supply parties on the levels above the Camp One spring. The rather unusual taste that resulted from treatment of water with halazone was in such contrast to the usually clear flavor of cold cave water that, despite instruction to the contrary, many persons drank the unpurified cave water on the sly and avoided the halazone water which was nicknamed "Wanger Water." When water containing halazone was used in the carbide lamps, the lamps clogged and would not light. Since drinking water and lamp water are usually carried in the same container on caving trips, the majority (but not the doctor) decided that the use of halazone was impractical. However, there have been no adverse effects from the use of regular cave water.

The disposal of human body waste presented another problem which was solved by digging a straddle trench about a quarter of a mile from Camp One. It was a good distance from the spring, so that there would be no danger of contamination. Chlorinated lime and a spade were brought into the cave for use in the trench. Garbage and trash were burned, and that which was not burnable was buried.

Because of the strenuous exertion, everyone perspired excessively and, due to high humidity, evaporation from skin and clothing was slight. An hexachloraphine soap was utilized exclusively for bathing because of its value in the destruction of bacteria, which lessened the hazard of infection. Fresh perspiration is odorless until odor-producing bacteria can begin their action.

III. EPILOGUE

After the expedition, a questionnaire was compiled which, when answered by each individual, would give some idea of the effect of a week underground both on the group as a whole and on each individual. All but two expedition members filled out a questionnaire. The following conclusions, some of which are tentative, have been made.

There were sixty-four persons on the expedition, thirteen of whom were women. The age range of the group was from nineteen to seventy-seven (nineteen to fifty-seven for those who went underground). Fifty persons, including five women, spent a total of 4,646 hours underground (an average of nearly ninety-three hours each), the periods ranging from eight to one hundred sixty-seven hours, or an average continuous period of seventy-seven hours per person. Forty-nine of the fifty persons who went underground returned questionnaires after the expedition. Thirteen of these claimed better health as a result of the experience, eight of the thirteen finding relief from colds or sinus conditions. Four, two of whom complained of fatigue, said they felt slightly worse afterwards, and the remaining thirty-two noticed no change in health.

For the forty-one who were in the cave long enough to require rest, the average sleep per twenty-four-hour period was 5.9 hours. Five persons had never used a sleeping bag before, whereas twenty-five had no previous experience in sleeping in a cave. Six stated that the quality of their rest was poor; two of these had never been in a sleeping bag before; the remainder claimed quality of sleep ranging from fair to excellent.

The group as a whole averaged three meals per day. The general consensus was that, while the food was adequate in quantity, the hours for eating were irregular and the type of food was poorly selected for consumption underground. Irregularity of meals, of course, was inevitable because of the unusual hours which were kept.

Seven persons complained of the low temperatures in the cave, and eight of the humidity. In reference to the absence of day and night inside the cave, seven individuals stated that they were confused as to whether it was morning, afternoon, or night, and two persons said that they missed daylight. Of the minor ailments suffered in the cave, fatigue led the list with thirty victims. Headaches, insomnia, and colds were second, with six instances of each. Shortness of breath and coughs claimed five victims each.

While the majority experienced no change in mood, five claimed irritability or frustration, and four stated they were in a better mood than usual. Dampness, cold, food, and fatigue led the list of dislikes of those underground, whereas exploration and companionship were far ahead of any other enjoyments of the trip.

Of the sixty-three persons who stated that they would go on another similar expedition, a number qualified their affirmative answers; only one said he would not attempt another such trip. He was a reporter; not an NSS member.

3

by Alden E. Snell

SUPPLY REPORT

THE PROBLEM OF SUPPLY for an undertaking of the magnitude of the Floyd Collins Crystal Cave Expedition was one of primary importance, requiring detailed advance planning. The equipment, supplies, and menu needs of the work parties living in the cave were completely different from those of the support personnel working from Base Camp. For example, while fresh meat and vegetables were served in the Base Camp mess hall, concentrated foods were sent to Camp One.

Many staple food items, as well as equipment, were donated to the expedition, so it was necessary to plan menus which would utilize these items. Daily trips to Cave City and to the town of Horse Cave to obtain supplies were necessary, and considerable savings were effected by buying in quantities through wholesale houses.

One of the unexpected obstacles snarling the supply line was the outbreak of forest fires in the area. With some exploration personnel already underground at the onset of the fires, the job of fire-fighting fell to the support personnel. This resulted in an inadequate supply of food in Camp One for the first few days, even though available supply teams made extra trips with insufficient rest to try to make up for lost time. The Gurnee cans, sheet-metal tubes pointed on one end and with a line attached to the nose to facilitate transport, were loaded with supplies and literally dragged through the Crawlway. They were sometimes attached to the individual's foot in particularly tight spots. The condition of the Gurnee cans as they were returned to the surface attested to the punishment they took hauling a ton and a half of supplies through the Crawlway.

261

Since drinking water was available at both Camp One and Camp Two, every effort was made to supply dried and dehydrated foods in order to lighten the burden of the supply teams. However, caving being strenuous work, calls were received from below for more substantial foods.

The supply teams were found to be most efficient when composed of three or four persons. Groups of all sizes, from two to ten or twelve, were tried, and the larger parties were definitely not as effective as the smaller, either in the time consumed for a round trip or in the poundage of supplies carried. The most efficient trips were made by a four-man team carrying not more than twenty-five pounds each.

Supply teams were started into the cave at any time personnel were properly rested and ready to go. Consequently, they arrived back at the Base Camp at all hours ready to be fed. This meant that a mess attendant had to be on duty at all times, and meals were available twenty-four hours a day. Trail lunches were also prepared by the mess hall for the supply teams to eat in the cave.

Some most necessary supplies were the most troublesome to carry. For instance, the amount of carbide needed by the advance parties was inordinately heavy to transport through the crawlways to Camp One and Camp Two. The hazard of transporting gasoline for underground cook-stoves was partially alleviated by bearers wearing electric head lamps, but the electric cable and the bulky battery on the belt required for these lamps made them extremely cumbersome in tight sections of the Crawlway. A fuel less volatile than gasoline is recommended for future expeditions of this nature.

Telephone communications between the one surface and two underground camps expedited supply in that the supply officer was informed immediately which underground stocks were depleted. Requests for extra clothing or luxuries, such as tobacco and candy, were complied with by the supply teams whenever possible, thus boosting morale below.

The majority of the personnel, all seasoned cave explorers, who made only one trip into the inner camps, complained of the ruggedness of the passageways, so supply-line personnel, who made an average of five round trips during the week (one man making nine), deserve special commendation.

4

by Robert L. Lutz

COMMUNICATIONS REPORT

THE ACCOMPANYING SCHEMATIC DRAWING shows the lay-out for the telephone system. A twelve-line Army switchboard of the magneto type and fourteen Signal Corps type EE8B telephones were used. The long wires were composed mostly of W110B Army field wire. A total of ten miles of wire was used, of which eight miles was laid in the cave. The EE8 Army telephones stood up exceedingly well under the very rough treatment they received in being dragged into the cave. It is remarkable that they worked at all. All functioned very well, however, with the exception of one telephone which was initially in poor condition. The ringing circuit on this telephone was inoperative due to dirty ringer contacts.

The outside line, which the expedition tapped into the fire line leading to the Mammoth Cave dial system, was isolated from the expedition's system by using an Army surplus C161 repeating coil with a capacitor in the tip side of the line. A commercial type dial telephone was bridged across the dial side of this outside line, and this line was connected to an auxiliary switchboard which was used to provide additional switchboard drops in the telephone system. The auxiliary switchboard was a converted range-officer's spotting board.

All circuits were simple metallic pairs except a phantom circuit that was used to Camp Two. The Camp Two drop at the switchboard was connected to the center taps of the two loading coils carrying the two Camp One circuits. This made it possible to take three circuits through the Crawlway on two metallic pairs. At the supply route junction where the Camp Two trail left the Camp One trail, a repeating coil was placed in each pair. The

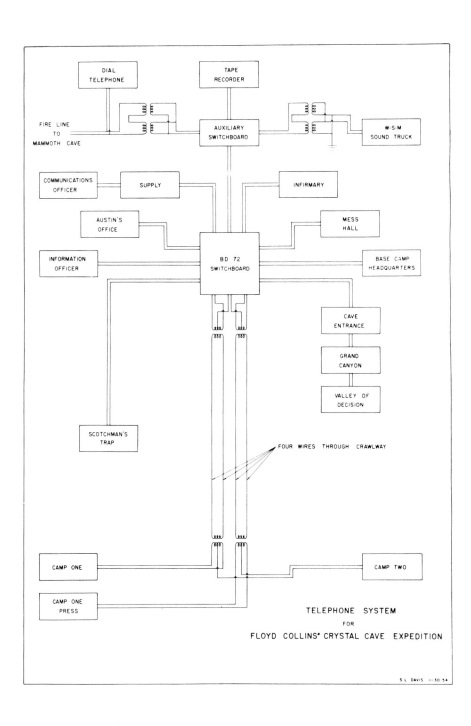

TELEPHONE SYSTEM

FOR

FLOYD COLLINS' CRYSTAL CAVE EXPEDITION

S L DAVIS 11-30-54

Camp Two circuit was taken from the center-taps of these repeating coils.

To conserve wire and switchboard drops, two party lines were used. The Valley of Decision, Grand Canyon, and the cave entrance were on one party line, while the other party line included the communication officer's telephone and the supply tent telephone.

Initially, the communications team tried using ground-return circuits for all telephones in the hope that they would have to lay only half the amount of wire for the same number of circuits. However, ground noise was so excessive, due to poor grounding and other factors, that ground-return circuits were abandoned in favor of metallic circuits providing much better sound quality. It was learned that wires should be laid well out of the pathway, out of water, and in the air as much as possible. Wires were transposed to eliminate cross-talk.

Some personnel received mild shocks from the ringing voltages. This was due to rather high leakage to ground, causing a potential difference between the handsets of the telephones and the ground upon which these people were standing. The use of a line with a slightly higher insulation resistance, and the isolating of the switchboard from the ground would have prevented this difficulty.

In the future, experiments should be carried out to determine the feasibility of using lightweight telephones and wire for communications in large cave systems.

by Thomas C. Barr, Jr.

GEOLOGICAL REPORT

THE GEOLOGICAL HISTORY of Floyd Collins' Crystal Cave begins in the Mississippian Period, 260 million years ago, when Kentucky was part of a shallow inland sea. Lime in the shells of molluscs and lime occurring as mineral matter in the sea water became deposited in a thick layer over the floor of this sea. Rivers from the land brought mud, sand, and gravel into the sea; these materials also were deposited from place to place, in rather definite layers.

When the sea receded, these sediments were compacted and solidified into rock. The limy ooze became limestone, the mud became shale, the sand became sandstone, and the gravel was transformed into conglomerate. Some of the shells of marine animals which had lived in the ancient Mississippian Sea were embedded in these rock strata and are found today as fossils. For millions of years the limestones in which Crystal Cave and Mammoth Cave have been developed lay buried beneath thick layers of relatively insoluble conglomerates and sandstones. Water could not reach the limestone to begin its slow work of cave making.

Once the sandstone cap had been removed, the limestone was subject to the action of water. Water can affect the limestone in two ways—one is by the simple, mechanical process of abrasion; the other is by solution. Lime, or calcium carbonate, is a relatively insoluble substance. Rainwater picks up carbon dioxide in the air to form a weak solution of carbonic acid:

$$H_2O \quad + \quad CO_2 \quad = \quad H_2CO_3$$
(Water and Carbon dioxide yield Carbonic acid)

This acid combines with calcium carbonate to produce soluble calcium *bi*-carbonate:

$$H_2CO_3 \quad + \quad CaCO_3 \quad = \quad Ca(HCO_3)_2$$
(*Carbonic acid and Calcium carbonate yield Calcium bicarbonate*)

Long after the Mississippian Sea had disappeared, a warping of the earth's crust pushed the rock strata upward into a great dome-like structure called the Cincinnati Arch. Erosion began to remove the upper layers of these rocks, cutting a huge basin in what is now southern Ohio and Central Kentucky. Since this erosion was most active at the highest part—the center of the arch—terraces of the younger, more resistant rocks were left at the edge of the dome structure. It is in such a terrace that Crystal Cave occurs.

The little town of Cave City lies on a comparatively flat, gently rolling plain known as the Pennyroyal Plateau. The Pennyroyal is floored with thick Mississippian limestones that are older than those in which Crystal Cave is formed. To the north and west of Cave City lies the Dripping Springs Escarpment; it is within this upland, a terraced remnant forming an escarpment at the edge of the Cincinnati Arch, that so much cavern development has taken place. Capping the escarpment is a thin layer of resistant sandstone, called the Cypress sandstone; beneath it are thick layers of limestone, sometimes called the "Mammoth Cave limestone," but more properly divided into two parts, the Renault-Paint Creek limestone and the Ste. Genevieve limestone. These rocks all dip to the northwest at about thirty feet per mile, toward the center of a great structural depression, the Western Kentucky Coal Basin.

Underground streams flowing beneath the surface of the Pennyroyal Plateau sink beneath the escarpment, their final destination a mystery. The Green River crosses the Plateau and has cut a deep canyon into the escarpment, but other surface streams are infrequent; most of the drainage is underground. Large springs issue from the southern wall of the Green River canyon. The most prominent of these are the outlet of Mammoth Cave's Echo River, the outlet of Styx River, also in Mammoth Cave, and Pike Spring, just below the entrance to Crystal Cave. Pike Spring is the largest of these springs, the stream issuing from a low, dark opening that extends only a hundred feet or so before the ceiling drops to the level of the water. These springs are probably the final outlet of some of the underground streams that sink beneath the escarpment. But no one knows where the water travels between its disappearance near Cave City and its final reappearance on Green River.

At a certain depth below the surface of the ground, there is a level beneath which all of the voids between the rocks are filled with water; this level is the *water table:* Above the water table rain water entering the ground percolates downward between the soil particles and through joints (natural crevices extending more or less vertically through the rock) and along bedding planes (zones of weakness between the layers of rock). This part of the ground is called the *vadose zone* (Latin, *vadosus,* "shallow"). Below the water table is the *phreatic zone* (Greek, *phreator,* "well"), where all of the voids are completely filled with water.

It was once thought that water in the phreatic zone was static, except for occasional fluctuations in the level of the water table. But now we know that there are definite channels of movement below the water table, and that these deep streams of water may perhaps move upward and downward as well as horizontally. In a limestone district they will follow the joints and bedding planes in the rock—just as the vadose water does, thus cutting their channels along the paths of least resistance.

One way in which a cave may be formed is by vadose water entering the ground through a conical depression called a *sinkhole.* This water soon cuts a narrow channel for itself, and may "capture" smaller trickles of water flowing through the limestone. It may take thousands of years before this embryonic cave reaches a size that can be explored by man. Such a cave would resemble a river system, with a main channel and tributaries, and would become smaller as one proceeded upstream, toward the source of the water.

But some caves have a rather peculiar pattern, consisting of loops, splits in the passage, and networks of passages. It has been suggested that most large caves and many small ones have been developed below the water table, where all the voids are filled with water, and the surface processes of stream development do not occur. That such waters can excavate a cave has been demonstrated by geologists of the Tennessee Valley Authority, who found openings hundreds of feet beneath a solid rock river bed while advising engineers about the nature of the rock in which dams were being constructed.

The bottom of X Pit, in Crystal Cave, lies near the water table, and in wet weather when the water table rises, it is inundated. The presence of eyeless fish and crayfish in the X Pit pools suggests a connection with the phreatic network of sub-water-table streams.

An important observation was made by engineers studying the Mammoth Cave area to determine whether or not it would be feasible to build a dam on Green River. They tested underground streams with fluorescein, a green dye that can be detected in minute concentrations. Some of this dye was poured into pools and little streams in Colossal Cave, on the opposite side of Flint Ridge from Crystal Cave. It was expected that the dye would reappear in Houchins Valley, between Flint Ridge and Mammoth Cave Ridge, below Colossal Cave. To the surprise of the engineers, the fluorescein colored the waters of Pike Spring. This indicates the existence of a drainage channel cutting northeast through the entire width of Flint Ridge. It is probable that this channel existed when Mammoth Cave Ridge and Flint Ridge were connected.

This kind of evidence has led geologists to suggest that many of the caverns in the Mammoth Cave and Crystal Cave area have been formed beneath the water table, in the zone of saturation. As a matter of fact, the confusing three-dimensional networks of passages in both Mammoth and Crystal are quite difficult to explain by processes of simple stream drainage. If cave development had taken place entirely in the vadose zone, one would expect the caves to resemble a drainage pattern of the form assumed by a river and its tributaries. But this is not the case at all. The map of Crystal Cave shows a series of parallel passages zigzagging back and forth above a big, lower level avenue (Floyd's Lost Passage) which runs in a different direction. Splits, loops, and blind alleys can hardly be attributed to vadose origin.

The concept of *levels* in the Mammoth Cave region needs to be carefully considered. Traditionally there are supposed to be five levels in Mammoth Cave, each level formed by vadose streams emptying into Green River when the water table was at the corresponding level. Through erosional processes the Green River deepened its channels and in so doing lowered the water table. If this mode of cavern origin is correct, we ought to find topographic indications of four pauses in the development of the surface in the cave region, and all of the other caves in the area should have levels that correspond to the five observed in Mammoth Cave. Many of the levels and passages in Mammoth do not fit precisely into one level or another, but vary in elevation from place to place. On the surface we look for evidence of the four pauses in the process of land erosion—flats and hanging-branch valleys tributary to Green River—and find no satisfactory correlation between surface levels and cave levels. If we tried to establish levels in Crystal Cave we would probably have between eight and ten, perhaps more.

After geologic uplift of the cave area has occurred, the cavern passages which have been excavated by the phreatic channels are drained of their water and left "high and dry." There are virtually no surface streams in the cave region, all of the surface water quickly finding its way underground into the spaces already carved out. These secondary streams usually leave traces of their work behind. In the Gypsum Route of Crystal Cave the ancient path of such a stream can be seen meandering back and forth across the passage at about shoulder height. This stream probably entered at a sinkhole that opened into the cave near Devil's Kitchen and flowed back into the avenue leading toward Scotchman's Trap. At that time the passage through which it flowed was largely filled with clay, which was later undercut by another stream and partially removed, so that the present floor is four and a half to five feet below the meander cuts. These stream meanders, or bends, occur with surprising regularity about thirty or forty feet apart, first on one side of the passage, then on the other.

Clay, sand, and gravel are deposited in the cave in different places. These deposits, or *fill*, may give us a clue to the origin of different parts of the cave. The finer materials, especially the clay, are more likely to be deposited by deep, slowly moving streams, such as those below the water table, while the larger particles, like the gravel, are more readily carried by faster-moving, free surface streams. Moreover, the amount of surface water would be greater in the winter and spring, and we might expect banding in the fill, because of thin layers laid down each wet season. Water coming into the cave from a surface opening would bring sand and gravel from the overlying rock strata, but we would expect primarily clay, derived from insoluble matter in the limestone, in deposits which were laid down by the phreatic water. In places in the Crawlway from Scotchman's Trap to Straddle Canyon, there is a one-foot-thick deposit of gravel on top of clay; this gravel is quite similar to the gravel seen on top of the Cypress sandstone above the cave, and seems to be derived from the Pottsville conglomerate, a formation that once overlay the cave region. Perhaps the stream that entered at Devil's Kitchen found its way to a lower passage and flowed along the Crawlway, depositing gravel on top of clay. It may have cut the deep trench down the middle of Straddle Canyon.

Much of Crystal Cave thus appears to have been excavated below the water table and later modified to a considerable extent by secondary streams after uplift had occurred. Probably all caves begin as embryonic networks of fissures below the water table, but a great deal of data will have to be accumulated before we can say how much of Crystal Cave is of phreatic origin and how much is due to the work of vadose streams.

When a cavern attains a certain size, collapse may occur, because of the weight of the overlying rock pressing down on the strata that form the ceiling. Usually such breakdown ceases when the passage roof acquires an arched shape, because an arch is mechanically stronger than a flat roof. Although cave explorers almost never see falling rocks, they are careful not to dislodge any loosely arranged piles of rock. In Crystal Cave, where the limestone beds lie relatively flat, most breakdown occurs as slabs that have fallen from the ceiling. In a few places large blocks have broken off from the walls.

A prominent feature of Crystal Cave is the large number of domepits. The term *domepit* is used, because a vertical cavity can be a pit if the observer is at the top, or a dome if he is at the bottom looking up. The largest domepits in Crystal Cave are X Pit, C-3 Waterfall, and the Bottomless Pit. Domepits may develop under both vadose and phreatic conditions. When the cave is still below the water table, solutional enlargement may take place along a vertical joint where a higher level passage crosses a lower one, but surface waters may also cut domepits in a cave in the vadose zone. The pit below Ebb and Flow Falls is very likely of vadose origin. Some domepits were begun by phreatic action and brought to their present stage by cutting action of vadose water.

Once the cavern passages have been elevated above the water table, the way is open for formation of *dripstone*. Where small trickles of water percolate down into the cave they spread out over the ceiling and walls and may collect into drops. On their way down they have accumulated a small amount of calcium bicarbonate in solution; when they reach the cave atmosphere, release of carbon dioxide and evaporation take place, and a tiny deposit of calcium carbonate is left behind.

As these deposits grow downward they resemble soda straws, with a cavity in the center. Finally the hole becomes plugged with calcite crystals and the formation—in this case a *stalactite*—can increase in diameter and length. Some of the drops may fall to the floor and begin building a formation that grows upward—a *stalagmite*. Stalagmites have no hole through the center, of course, but there may be a splash cup on top, where the water droplets hit the stalagmite. A cross-section cut through a stalactite will show the central cavity and layers of calcite arranged concentrically around it, looking very much like the growth rings of a tree trunk. These layers indicate alternating periods of wet and dry weather, and frequently correspond to years. The rate of growth of stalactites and stalagmites varies considerably, depending upon the composition of the limestone, the amount of ground-water entering the cave, the relative importance of evaporation and release of carbon dioxide, and the surface area of limestone to which the ground-water has been exposed. Deposition is at best quite slow but probably somewhat faster than the oft-quoted "average" of a hundred years per cubic inch.

The sandstone cap and the depth of most of the passages in Crystal Cave account for the scarcity of dripstone, since what little ground-water finds its way into the limestone cannot reach most parts of the cave without being captured by small underground streams. Because ground-water travels through the joints, dripstone is most common near or along open joints in the rock. When water flows slowly over the walls in a sheet, a mass of *flow-stone* may be deposited. When water collects in pools with a gentle over-flow, part of the overflow may evaporate, forming *rim-stone terraces* around the edges of the pool.

The direction of growth of stalactites, stalagmites and the common forms of dripstone is largely controlled by gravity. But in the Helictite Route (and in a few other parts of the cave, as in Mud Avenue) there is a remarkable display of a kind of dripstone that seems to defy gravity. These formations —*helictites*—begin as stalactites but soon turn up, down, or sidewise in a most peculiar manner, sending out many branches. Why this occurs is not completely understood. Perhaps it is due to the capillary action of water, a thin film spreading out over the surface of the formation. Air currents were once believed to cause this anomalous structure. The most recent theory is that the direction of growth of the crystals of calcium carbonate causes the eccentricity—one layer of crystals will turn slightly to one side, the next layer building on top of the first, oriented still further in the direction of twisting.

By far the most unusual of the formations in Crystal Cave are the numerous gypsum deposits. These were formerly much more numerous than at present, many of them having been broken off since the cave was dis-covered. Gypsum is calcium sulfate. The source of the sulfate has not been determined. Gypsum in American caves seems to be most abundant in lime-stone of comparable age in Tennessee, Kentucky, and Alabama.

Cave gypsum occurs in a fibrous form called *selenite*. It may encrust the walls in rosettes, coat the ceiling, or accumulate on the floor as gypsum sand. A rare form occurs as masses of tiny needles that sway back and forth at the slightest breeze. The most spectacular type of gypsum resembles flowers, and a special name—*oulopholites*—has been given to this variety. Oulopholites appear in the Gypsum Route and at one end of Floyd's Lost Passage, where the white selenite crystals look as if they had been squeezed out of the walls and ceiling. Like the helictites, oulopholites pose a problem to mineralogists, because they are difficult to explain. The crystal fibers seem to grow outward as they form at the base of the oulopholites.

Various other formations are encountered in Crystal Cave. One of the most unusual of these is *moon-milk*, a semi-liquid paste of precipitated chalk, which was found beneath loosely-attached "grape cluster" (*botryoidal calcite*) formations in Floyd's Lost Passage. The grape cluster consists of little spheres of calcite on the end of short stalks. It is probably laid down by a capillary sheet of lime-laden water spreading out over the walls of the cave.

The geological study of caves is still in its infancy. There is much to be learned and there are many theories to be tested. Because of its size and its location in the heart of Kentucky's cave district, Crystal Cave offers much to the cave geologist, but it is only one link in a great chain. Only by a careful investigation of the features of many caves can we obtain a comprehensive knowledge of the processes by which caves originate and by which their fascinating features develop.

Suggested Technical References

Bretz, J Harlen (1942). "Vadose and phreatic features of limestone caverns." *Journ. Geol.* vol. 50, No. 6, pt. 2, pp. 675-811.

Davis, W. M. (1930). "Origin of limestone caverns." *Bull. Geol. Soc. Amer.,* vol. 41, pp. 475-628.

Livesay, Ann (1953). "Geology of the Mammoth Cave National Park Area." *Kentucky Geol. Surv.,* Spec. Publ. No. 2, Series 9, 40 pp.

Lobeck, A. K. (1928). "The geology and physiography of the Mammoth Cave National Park." *Kentucky Geol. Surv.,* Series 6, vol. 31, pt. 5, pp. 331-339.

Moore, G. W. (1953). "The origin of helictites." *Occ. Papers of the Nat. Speleological Soc.,* No. 1, 16 pp.

Swinnerton, A. C. (1942). "Hydrology of limestone terranes"; in *Physics of the Earth*—IX, Hydrology, pp. 656-677, McGraw-Hill Book Co., New York, N. Y.

Weller, J. M. (1927). "Geology of Edmonson County, Kentucky." *Kentucky Geol. Survey,* Series 6, vol. 28, 246 pp.

by Donald N. Cournoyer

METEOROLOGICAL REPORT

THE FLOYD COLLINS' CRYSTAL CAVE EXPEDITION afforded an opportunity to study the underground meteorological conditions of an extensive cave system for a continuous period of 168 hours. It was originally planned to set up a weather station on the surface near the cave entrance and another in Floyd's Lost Passage in order to determine the relationship between inside and outside atmospheric conditions. Unfortunately, enough instruments were not available for the surface station. However, data collected from the cooperative U. S. Weather Bureau station at Mammoth Cave was utilized for this purpose.

METEOROLOGICAL INSTRUMENTS

The following instruments were used in observing the meteorological conditions in the cave.

Instrument	To Measure:
Thermometer	Temperature
Psychrometer	Humidity
Anemometer	Wind velocity
Staff Gage	Water level fluctuations

The thermometer consists essentially of some confined substance, such as mercury, which changes volume with a change in temperature. A sling psychrometer consisting of wet and dry bulb thermometers is whirled by the observer to obtain the depression of the wet bulb, from which is computed the humidity or amount of moisture content in the air, as well as the dew point. The anemometer consists of twenty wind vanes which rotate clockwise by the force of air currents. It registers the velocity of the air current on a recording dial fixed in the center of the wind vanes. The staff gage is a graduated scale fixed in the stream or pool and is used in measuring fluctuations of water level.

A thermograph and a barograph were taken into the cave, but the Gurnee can containing these instruments had to be abandoned in the Crawlway due to fatigue of the party carrying it. These instruments were recovered by William T. Austin on February 17th, 1954. The instruments had received too much abuse on the way through the Crawlway to be serviceable. The meteorologist attempted to overhaul the barograph by using parts of the thermograph linkage. However, the overhaul was not successful, so data collected on the barograph were not usable.

CLIMATIC SUMMARY OF THE REGION

The general cave area lies within the path of the moisture-bearing low-pressure areas that move from the western Gulf region over the Mississippi and Ohio Valleys to the Great Lakes and the Northern Atlantic Coast. There is, consequently, a great variation in climatic elements.

The average annual precipitation in the Floyd Collins Crystal Cave area is three to six inches greater than the general average of forty-six inches for Kentucky. The Weather Bureau lists the mean annual temperature for this area as 58.8°F. Average daily temperature in the Floyd Collins Crystal Cave area for the month of February is 38°F. with an average daily range of 18°F.

TEMPERATURE

A short distance beneath the surface of the earth the temperature fluctuates very little, and in most cases approximates the mean annual temperature of the area. In many caves the air temperature coincides with the mean annual temperature of their location. However, in Floyd Collins' Crystal Cave the average temperature was 54°F. but the mean annual temperature of the surface area is 58.8°F.

Cave-temperature records collected throughout the period of the expedition indicated that the temperature is generally constant from one location to another although it may vary as much as plus or minus 3° or 4°F.

Two thermometers were set up in Floyd's Lost Passage close to an opening that leads into Mud Avenue. Both were read at intervals to determine whether or not they agreed with each other. Every effort was made to screen out the effect of body heat upon the temperature elements so that the data would be reliable. Temperature and humidity in Floyd's Lost Passage and in Grand Canyon Avenue were recorded and are shown in the table on page 278.

HUMIDITY

The humidity was measured by the use of the sling psychrometer to obtain wet and dry bulb temperatures. This instrument was whirled several times until two successive readings of the wet bulb agreed. The humidity remained fairly constant and was recorded as 97 percent, except for a sudden drop to 78 percent, when a drop of 9°F. in temperature occurred.

METEOROLOGICAL OBSERVATIONS IN THE CAVE

On the evening of February 15th, 1954, at 2300 hours, a location near the gypsum flowers in Floyd's Lost Passage was investigated for a temperature difference as it had been reported that the temperature here seemed higher than elsewhere in the cave. Checking at intervals by several speleologists revealed a constant temperature of 59°F., which was 5°F. higher than the general temperature observed in the cave. It is possible that, since this part of the passage is dome-shaped, the air currents tend to leave warmer air entrapped at this spot.

At the north end of Floyd's Lost Passage, another entrance is thought to exist. Cave rats have been seen here as have their nests, which consist of debris brought in from the surface. Air currents were noted coming from this general direction. The temperature at 0130 hours February 16th, 1954, was observed by Audrey E. Blakesley and the author to be 52°F. A trace of air current could be picked up by the anemometer, but it was not sufficient for a reading to be registered. On February 18th, 1954, at 1400 hours, a direct reading of wind velocity was registered at 2 to 3 mph, at the north end of Floyd's Lost Passage. The temperature here was 3°F. lower than elsewhere in the cave for this day.

It was also here in 1949 that James W. Dyer, Luther Miller and William T. Austin experienced a change of direction in air movement and a change in air temperature. In an attempt to determine where and if the air was reaching the surface, an automobile tire was burned to create smoke. The experiment, however, was not successful, as no smoke was seen at the surface. The reason could have been that the same phenomenon of reversal of air currents observed by the 1954 expedition had taken place.

On February 16th, 1954, at 0330 hours, the author and two other speleologists, who were sitting in the junction between Floyd's Lost Passage and Mud Avenue, felt a strong current of cold air. The thermometers revealed a sharp decline in temperature of 9.5°F., from 54.5°F. to 45°F. Probably the warmer air in the cave was being displaced by colder air from the surface and hence this colder air temporarily modified the temperature. The cold air was advancing so rapidly that it had not been warmed up to the temperature of the surrounding cave rocks.

At 0400 hours a check on the drinking spring, northeast of Camp One, revealed the water temperature to be 50°F., whereas prior to this it had been 52°F. The change in water temperature was probably due to colder precipitation finding its way into the cave water system. A phone call to the Base Camp communication tent verified the change in weather conditions at the surface, as suspected. This weather change probably accounted for the change in air temperature and movement.

On February 17th, 1954, at approximately 1100 hours, accumulated trash was burned about 500 feet northeast of Camp One. At the same time several explorers observed the candle flames and noticed that the air currents were moving in a northeast direction away from the camp. In a matter of hours the air currents had reversed, filling Floyd's Lost Passage with smoke. Much of this smoke was dissipated from Floyd's Lost Passage into Mud Avenue. A surveying party observed the smoke in the Flat Room, and about the same time a tourist guide saw smoke in the Helictite Route, the cave's highest level.

HYDROLOGY

In the limestone area of Floyd Collins' Crystal Cave no permanent surface drainage exists other than the Green River. Precipitation falling on the area sinks into the ground and ultimately finds its way to the Green River.

A system of underground streams is believed to exist in the lower depths of the cave, and the sound of running water could be heard coming from below. Attempts to get into these passageways were unsuccessful because of the difficult climbing and small openings.

A fluorescein test was conducted in an effort to discover a connection between a stream on an upper level and Bogardus Waterfall. The dyed water was not seen at Bogardus Waterfall; however, the test is inconclusive since it was not possible to continue observations at the waterfall a sufficiently long period of time.

Camp One was near an underground spring which was the expedition's water supply. A staff gage was set up at this spring and the water level was

METEOROLOGICAL DATA

Date 1954	Time Hours	Air temperature Degrees F.	Humidity Percent	Wind speed mph	Location
2/12	1500	56	98	**6	Grand Canyon
2/12	2230	56.3	98	**7	" "
2/13	0730	*58	96		Collins' Casket
2/13	1330	56	98	**6	Valley of Decision
2/15	1700	55	97		Camp One
	2100	54.5	97		" "
2/16	0030	54.5	97	Trace	" "
	0330	45	78	"	" "
	0600	54	84	"	" "
	1200	54	88	"	" "
	1500	55	94		" "
	1800	54.5	94		" "
	2100	54	97		" "
2/17	0330	54	98		" "
	0600	54	97		" "
	1500	54	97	Trace	" "
	1800	54	98		" "
	2100	54.5	96		" "
2/18	0000	54	97		" "
	0300	54	97		" "
	0600	54	97		" "
	1330	54	97	Trace	" "
	1600	54	97		" "
2/15	2300	59	100		Near gypsum flowers
2/17	1400	59	100		in Floyd's Lost Passage
2/16	0130	52	84	Trace	North end Floyd's
2/18	1400	51	86	2-3 mph	Lost Passage

Date 1954	Time Hours	Water temperature		Wind speed mph	Location
2/15	1800	52		Trace	Drinking spring
2/16	0610	50		"	" "
2/16	1810	52		"	" "

* Increase due to body temperature of personnel sleeping nearby.
** Speed of air moving out of cave measured at entrance.

Comparative precipitation data from U. S. Weather Bureau Cooperative Station at Mammoth Cave National Park, Kentucky:

		(*Precipitation in inches*)
2/15	Monday	Trace
2/16	Tuesday	0.96
2/20	Saturday	0.62

observed every three hours. On February 16th, 1954, the water level rose three quarters of an inch during the three hours between observations. Precipitation of 0.96 inch on the surface the same day was undoubtedly the reason for this rise. Increased dripping from the cave ceiling was noted, and some places became slightly muddier than usual. When the slight rise in water level was noted, the water temperature had dropped from the normal of 52°F. to 50°F. About twelve hours later the temperature was back to normal. By the afternoon of February 18th the spring water level was back to normal.

by Thomas C. Barr, Jr.

BIOLOGICAL REPORT

T HE EXISTENCE OF LIFE in total darkness has always intrigued the human mind. We receive eighty percent of our impressions from our sense of sight, and when we are deprived of the use of our eyes we find ourselves in serious difficulty. To many animals, however, the possession of eyes is superfluous. Absence of eyes is a normal feature of many of the so-called *cryptozoic* animals—those which spend all of their time beneath stones or in the soil—and many deep-sea animals. Therefore, cave animals that have lost their eyes and body pigment are not unique in this respect.

In Floyd Collins' Crystal Cave life is scarce. This is primarily a result of the difficulty of finding food within the cave. On the surface, animals depend on green plants, which manufacture food from water and carbon dioxide, utilizing the energy of sunlight. Fungi and bacteria are the only plants that can grow in total darkness, but they normally require some kind of organic matter on which to maintain themselves. One investigation in the Mammoth Cave area indicates that the spores of most species of cave fungi may be sterile. There are probably no species of plants indigenous to caves; the kinds of plants that do develop there germinate from spores brought in by air currents or various animals, including man.

Total darkness is the most constant factor of the cave environment. In most caves there is a dimly lit transitional area, the *twilight zone,* near the entrance. There is no true twilight zone in Crystal Cave, which has no known natural entrance except for narrow cracks and crevices communicating with the surface. A high relative humidity is another important factor in a cave. In many parts of Crystal Cave the humidity approaches 100 percent. Cave-dwelling animals are thus not concerned with loss of water from the body through evaporation. In most parts of the cave the temperature does not vary more than one or two degrees from 56°F. throughout the year. Rain-water entering the cave as small streams, and air currents from crevices opening to the surface produce the most noticeable fluctuations from this average temperature. But even the larger temperature changes observed in Floyd's Lost Passage are minor ones when compared to the seasonal temperature range on the surface.

Animals that show special characteristics associated with their life in caves—such as eyelessness and loss of body pigment—are known as *troglobites*. They could not live under conditions unlike those in the cave. Another group of animals living their lives underground are the *troglophiles*. They have life requirements that are readily met by the cave environment but have not evolved troglobite characteristics. *Trogloxenes* frequent caves and spend much of their time underground but must return to the surface at some time during their life cycle, usually for food.

The trogloxenes are important in the economy of most cave life communities because they bring food into the cave. Common trogloxenes are bats and pack rats. The excreta of these animals, plus the green leaves and other organic debris which the rats drag to their nests, provide nutriment for many smaller species of cave animals. Bats and pack rats are scarce in Crystal Cave, and their contribution toward supplying other animals with food must be small. Streams of water and air currents are in this case probably much more important food-carrying agents.

The little pipistrelle (*Pipistrellus subflavus*) was seen in Floyd's Passage. Only five individuals of this bat were counted during the week underground. A big brown bat (*Eptesicus fuscus*) was found hibernating near Grand Canyon in the commercial part of the cave. A pack rat (*Neotoma magister*) was seen near Floyd's Kitchen. Near Ebb and Flow Falls there was an abandoned pack rat nest, made of the fibers of an early explorer's rope, carefully unravelled and coiled and padded on the inside of the nest.

The only true troglophile found in the cave is the cave cricket (*Hadenoecus subterraneus*). It is not a true cricket and is incapable of even the slightest chirp, but it is a first cousin to grasshoppers and the true crickets. Although rather common throughout the cave, it is not abundant. The species is not limited to the Mammoth Cave region, as are the troglobites from Crystal Cave, but also occurs in southern Indiana and the Cumberland Plateau of Tennessee and Alabama.

The blind cave beetle (*Neathanaeops tellkampfii*) was the most abundant species of any animal encountered in the cave. In Floyd's Passage it inhabits sand flats near the walls. These beetles are troglobites. Little more than a quarter of an inch long, they were seen scurrying about in many a crawlway.

Caves sometimes serve as places of refuge for animals that formerly inhabited a region, but have largely died out there because of changing climatic conditions. The blind cave harvestman, or "daddy-long-legs" (*Phalangodes armata*) is a representative of just such a group. About twenty species of the family to which it belongs linger on in temperate North America, living in caves or under rotting logs and leaf mold, but its close relatives (members of the same suborder) abound in the tropics. Two females and a male of this little cousin of the spiders and scorpions were collected beneath a piece of wood in a wet, seldom-visited part of the cave near Grand Canyon. *Phalangodes* is carnivorous, feeding upon other smaller cave animals.

One of the small streams running into Floyd's Passage was the home of a blind flatworm (cf. *Sphalloplana percaeca*). Several dozen of these tiny worms were observed crawling about in the shallow water of this stream, sharing their environment with a blind water louse, a small crustacean known as *Asellus stygia*. The spring that supplied drinking water to the explorers at Camp One was inhabited by a white, eyeless crayfish (*Orconectes pellucidus*). The group that descended into X Pit in search of the blind fish, *Typhlichthys*, had to be content with capturing a large crayfish of the same species.

Other troglobites observed in Crystal Cave were a little white spider (*Anthrobia mammouthia*), a white millipede (cf. *Scoterpes copei*), and a blind bristletail (*Campodea cooki*), a little insect related to the silverfish.

In many parts of the cave, especially in Floyd's Passage and the Bottomless Pit Trail, small flies (*Scoliocentra defessa*) were seen crawling about on the ceiling. This species, a trogloxene with large, well-developed eyes, is commonly found in caves in many parts of the United States.

It is generally assumed that eyeless, colorless cave animals have evolved from ancestors that had eyes and body pigment. Three principal theories have been developed to account for this change.

The first of these is the Lamarckist theory, which states that the eyes of each generation of a cave-dwelling animal become a little bit smaller, and each generation becomes a bit paler, until the cumulative effect produces a completely eyeless, colorless race of animals. This theory has been criticized because it implies that changes in the body of an animal during its lifetime are inherited by its offspring. If a mouse's tail is chopped off, its offspring will not have shortened tails, but if two mice born with short tails are mated, their offspring will have short tails. It appears highly improbable that the hereditary material, or genes, of an animal are changed by what happens to the body of the animal during its lifetime.

Biologists who applied Charles Darwin's theory of natural selection to the evolution of cave fauna suggested that the pale, eyeless animals were better suited to their environments than animals with pigment and eyes. They would therefore have a better chance of surviving in the struggle for existence and of reproduction of their kind. However, there soon arose the difficulty of explaining why the lack of eyes and loss of pigment would be especially advantageous in a cave.

Darwin himself did not subscribe to so strict an application of his own theory; he made a distinction between loss by disuse, as in lack of eyes, and true adaptations, such as longer antennae to compensate for loss of sight. Longer antennae would be true adaptations because they are distinctly advantageous to the animal, whereas the presence or absence of eyes would seem to be immaterial to an animal living in complete darkness.

There is a third theory which maintains that eyeless and albino mutations are neither advantageous nor disadvantageous but are neutral in adaptive value. An albino crayfish living in surface waters would be readily seen and captured by its enemies, so albinism would be disadvantageous above ground. But in the absence of light, the color of the crayfish might be immaterial to its survival. It has been shown that the same mutations tend to occur again and again in a given species, with a constant frequency for each mutation. Normally, albino mutations would be weeded out by the process of natural selection, but in a population of cave-dwelling crayfish, they could accumulate slowly until the whole population had become albinistic. Although this theory appears to be rather straightforward, it is in any case quite difficult to determine with absolute certainty whether a characteristic is advantageous or not, because some factor in the animal's biological make-up may have heretofore escaped our attention.

Suggested Technical References

Bailey, Vernon (1933). "Cave Life of Kentucky." *American Midland Naturalist,* Vol. XIV, No. 5, pp. 385-635.

Packard, A. S. (1886). "The cave fauna of North America." *Memoirs of the National Academy of Sciences,* 1886. 156 pp. xxvii pl.

INDEX

References to photographs appearing in the text are in bold type.

accidents in caves xxiii, 26-27, 87, 190, 193, 208, 254-255; *see also* safety in caves
altimeter 222-223, **223**, 224
American Association for the Advancement of Science xv
American Psychological Association 256
animal life in caves 158-159, **159**, **160**, 162-165, **166**, 268, 280-283; *see also* bats
Appalachian Apple Service, Inc. 255
archeological evidence 236, **237**
Argosy 100
Austin, Jacqueline F. xix, 43, 174, 227
Austin, William T. ix, x, xi, xix, 43, 65, 66, 85, 230, 275, 276; encouraging exploration in FCCC, xv; construction of new entrance, xv; taking Joe Lawrence into FCCC, 3-7, 11-20; description of, 5; photographs by, 12; as FCCC explorer and manager, 21, 149, 235; as assistant expedition leader, 32, 73-82, 90, 112, 167-169, 188, 203, 206; Mud Avenue trip, 171-175, **174**

B Trail 21, 96, 120, 144-155, 154 (map), 223; description of, 149
Barr, Thomas C., Jr., 35, 46, 94, 95, 117, 118-120, 156-166, **166**, **237**; photographs by, 166; geological report, 266-273; biological report, 280-283
bats 120, 164-165, 281
bibliographies of technical references 273 (geological), 283 (biological)
Bishop, Stephen xix
Black Avenue xv
Blakesley, Audrey E. 48, 117, 120, 131, 195, 224, 276; Fool's Dome trip, 133-142; rest trip, 206, 220-221, **221**; dye at Devil's Kitchen, 231-232
"Bless 'Em All" 227-229
Bloch, M. Girard 29-30

Bogardus Waterfall Trail vii, 18, 96, 120, 148-155, 223, 224, 230-232, 234, 238; Bogus Bogardus, xv
Bögli, Dr. Alfred xv
Bottomless Pit: survey to, xi, 88-89, 90-91, 110-113, 176-177, 191; description of, 63-64, 271
Bruce, C. N., photographs by 77, 123

Brucker, Ellen xix
Brucker, Roger W. vii, 20, 32, 39, 40, 81, **115**, 126, 129-130, 176, 239; writing of *The Caves Beyond*, xi-xiv, xxvii; wife, xv, xix; president of CRF, xvi; arrival before expedition, 43-46; on first expedition trip into FCCC, 51-72; trip in with Dyer, Miller, Harsham, Halmi, 82, 97-117; Fool's Dome trip, 133-142, **137**, **139**, **140**; Bogardus trip, 148-155; Flat Room trip, 177-181; Lost Passage trip, 181-185; Lost Paradise trip, 190-199; rest trip, 206, 220-221, **221**; elevation study, 222-223, **223**, 224-225; returns to Camp One, 224-225; trip to look for dye, 230-232; final trip out, 240-242
Brucker, Thomas A. xvii, xix
Brunton compass *see* compass, Brunton

C-3 Waterfall (Camp Two) 205, 206, 207 (map), 208, 213, **214**, **215**, **219**, 226, 237-238, 271
Camp Two *see* C-3 Waterfall
camping underground 15-19, 33, 37-38
canyon straddle (hopping), 59, 104, **106**
carbide lamp 39-40, 71, 97-98, 198, 258
Cascade Hall (Mammoth Cave) xvii
Cave City 4, 23, 188, 261; location of, 6, 267; hotel, 46, 47
cave, formation of 12-13, 165, 184, 233-234, 266-270
Cave Research Foundation (CRF) xvi, xix
Central Ohio Grotto xv, xvi
Charlton, Royce E., Jr. xi, 126, 133, 192, 199, **229**, **242**; first trip into FCCC, 4, 16-20; description of, 4, 32; trip to establish Camp One, 78-80, 93-96; Bogardus trip, 148-155, 234; Mud Avenue trip, 170-171; Flat Room survey, 177-181; Lost Passage trip, 181-185; at Camp Two, 209, 210-219
climbing equipment **41**, 42, 133, 140-141, 191
Coleman gasoline stoves 33, **227**, 262
Collins, Floyd viii, xix, 109-110, discovery of FCCC, 5; casket of, 7, 10; description of, as explorer, 7, 58, 69-70; death in Sand Cave, 8-10, 193; legends about, 23; home, **25**

285